Environmental Engineering

DESIGNING A SUSTAINABLE FUTURE

GREEN TECHNOLOGY

Environmental Engineering

DESIGNING A SUSTAINABLE FUTURE

Anne Maczulak, Ph.D.

An imprint of Infobase Publishing

ENVIRONMENTAL ENGINEERING: Designing a Sustainable Future

Facts On File, Inc.
An imprint of Infobase Publishing
132 West 31st Street
New York NY 10001

Library of Congress Cataloging-in-Publication Data
Maczulak, Anne E. (Anne Elizabeth), 1954–
    Environmental engineering : designing a sustainable future / Anne Maczulak.
        p. cm.—(Green technology)
    Includes bibliographical references and index.
    ISBN-13: 978-0-8160-7200-2 (alk. paper)
    ISBN-10: 0-8160-7200-0 (alk. paper)
    1. Environmental engineering. 2. Environmental protection. I. Title.

    TA170.M36 2009
    628—dc22   2009005030

Facts On File books are available at special discounts when purchased in bulk quantities for businesses, associations, institutions, or sales promotions. Please call our Special Sales Department in New York at (212) 967-8800 or (800) 322-8755.

You can find Facts On File on the World Wide Web at http://www.factsonfile.com

Text design by James Scotto-Lavino
Illustrations by Bobbi McCutcheon
Photo research by Elizabeth H. Oakes
Composition by Hermitage Publishing Services
Cover printed by Bang Printing, Brainerd, MN
Book printed and bound by Bang Printing, Brainerd, MN
Date printed: November 16, 2009
Printed in the United States of America

10 9 8 7 6 5 4 3 2 1

This book is printed on acid-free paper.

Contents

Preface

The first Earth Day took place on April 22, 1970, and occurred mainly because a handful of farsighted people understood the damage being inflicted daily on the environment. They understood also that natural resources do not last forever. An increasing rate of environmental disasters, hazardous waste spills, and wholesale destruction of forests, clean water, and other resources convinced Earth Day's founders that saving the environment would require a determined effort from scientists and nonscientists alike. Environmental science thus traces its birth to the early 1970s.

Environmental scientists at first had a hard time convincing the world of oncoming calamity. Small daily changes to the environment are more difficult to see than single explosive events. As it happened the environment was being assaulted by both small damages and huge disasters. The public and its leaders could not ignore festering waste dumps, illnesses caused by *pollution,* or stretches of land no longer able to sustain life. Environmental laws began to take shape in the decade following the first Earth Day. With them, environmental science grew from a curiosity to a specialty taught in hundreds of universities.

The condition of the environment is constantly changing, but almost all scientists now agree it is not changing for the good. They agree on one other thing as well: Human activities are the major reason for the incredible harm dealt to the environment in the last 100 years. Some of these changes cannot be reversed. Environmental scientists therefore split their energies in addressing three aspects of ecology: cleaning up the damage already done to the earth, changing current uses of natural resources, and developing new technologies to conserve Earth's remaining natural resources. These objectives are part of the green movement. When new technologies are invented to fulfill the objectives, they can collectively be called green technology. Green Technology is a multivolume set that explores new methods for repairing and restoring the environment. The set covers a broad range of subjects as indicated by the following titles of each book:

- *Cleaning Up the Environment*
- *Waste Treatment*
- *Biodiversity*
- *Conservation*
- *Pollution*
- *Sustainability*
- *Environmental Engineering*
- *Renewable Energy*

Each volume gives brief historical background on the subject and current technologies. New technologies in environmental science are the focus of the remainder of each volume. Some green technologies are more theoretical than real, and their use is far in the future. Other green technologies have moved into the mainstream of life in this country. Recycling, alternative energies, energy buildings, and biotechnology are examples of green technologies in use today.

This set of books does not ignore the importance of local efforts by ordinary citizens to preserve the environment. It explains also the role played by large international organizations in getting different countries and cultures to find common ground for using natural resources. Green Technology is therefore part science and part social study. As a biologist, I am encouraged by the innovative science that is directed toward rescuing the environment from further damage. One goal of this set is to explain the scientific opportunities available for students in environmental studies. I am also encouraged by the dedication of environmental organizations, but I recognize the challenges that must still be overcome to halt further destruction of the environment. Readers of this book will also identify many challenges of technology and within society for preserving Earth. Perhaps this book will give students inspiration to put their unique talents toward cleaning up the environment.

Acknowledgments

I would like to thank the people who made this book possible. Appreciation goes to Bobbi McCutcheon who helped turn my unrefined and theoretical ideas into clear, straightforward illustrations. Thanks also go to Elizabeth Oakes for providing photographs that recount the past and the present of environmental technology. I thank Jackie Cahi of the Kufunda Learning Village in Zimbabwe for providing information on building a sustainable community. My thanks also go to Marilyn Makepeace, who provided support and balance to my writing life, and Jodie Rhodes, who is a constant source of encouragement. Finally, I thank Frank Darmstadt, executive editor, and the editorial staff at Facts On File for all their help.

Introduction

Sustainability refers to the ability of a system to survive. It can be achieved if the majority of people work together in large ways and small. Energy production, transportation, construction, and other industries have large responsibilities to find methods for using resources in a sustainable manner. These present more complex challenges than recycling grocery bags or composting wastes, but the sustainability that will make a significant difference to the environment will likely come from major engineering projects.

Civil engineering has for centuries played a vital role in creating safe and functional structures for society. The welfare of a threatened environment now needs another component, sustainability, to complement safety and function. For this reason, civil engineering has given birth to the more specialized field of environmental engineering.

Environmental engineering combines all of the classic principles of engineering into a newer philosophy in which humans work with nature rather than try to invent ways to force their will on nature. Is it an exaggeration to say that humans have had a history of forcing nature into unnatural conditions? Some of civilization's greatest engineering feats have had tremendous impacts on the environment, either by altering habitat or interfering with the normal behavior and propagation of plants and animals. Even a beginning student in environmental science can recognize the damage done to the landscape by things such as the Great Wall of China, the Panama Canal, cross-continent superhighways and railroads, or the Alaska pipeline. These structures and similar engineering triumphs are true accomplishments that attest to the power of technology and innovation. For that, they have been valuable models for later engineering projects. Environmental scientists have learned, however, that such large engineering projects also have consequences that can be ecologically damaging. Environmental engineering has grown since the 20th century by replacing the older style of engineering with a new style that makes ecology a priority.

Environmental engineering's future seems boundless because it is based on the myriad ways in which nature solves its own engineering challenges. Nature does not use mechanical contraptions to move water uphill; it does not rely on *combustion* engines to make things go; and it does not require millions of miles of electrical power lines. Nature uses the materials that the Earth supplies to devise energy-generating systems, communication, mobility, and temperature and light sensors. Nature does all this with a minimum amount of energy input and produces waste that is 100 percent recyclable. People have yet to design a system that pumps water 200 feet (61 m) straight up toward the sky in a system that is silent, requires no mechanical pumps, and never malfunctions, yet giant sequoia trees do this every day. Clearly, environmental engineering has a distance to go to mimic nature's activities, but, fortunately, nature provides endless examples of processes such as the sequoias' that maximize energy conservation.

This book examines the emerging profession of environmental engineering. It discusses the ways in which environmental engineering blends the best aspects of art and design with the sciences of physics, geology, ecology, and the chemistry of matter. It begins with an overview of how environmental engineering grew out of civil engineering and now explores new arenas such as *ecological design, zero energy* architecture, and the concept of using nature's processes as a blueprint for human needs, an area called *biomimicry.*

The second chapter takes a close look at new transit systems that will achieve sustainability by using alternative fuel sources and also by offering communities new choices in travel. This chapter explores the various choices that people already have for getting places without using their personal vehicles, and it discusses the improvements all modes of travel will need to make in order to change people's reliance on cars. It covers public transit on rails and buses in addition to bicycles, air travel, and shipping.

Chapter 3 focuses on personal vehicles, perhaps today's most nagging cause of pollution, congestion, and habitat destruction (due to pollution and road construction). The chapter works on the assumption that personal vehicles will never disappear from society, so engineers must find ways to make vehicles much more fuel- and energy-efficient than they are at present. This chapter covers innovations in vehicle design for improving aerodynamics and power. It provides insight into new technologies in car shape, resistance, and surface technology.

Chapter 4 examines concepts in sustainable manufacturing. Energy-efficient and low-waste manufacturing have been built in very few places in the world, yet industry offers an enormous opportunity for countries to greatly reduce their total energy consumption. This chapter discusses emissions control and other waste control and the principles of reusing heat energy. It also ends with a discussion on the feasibility of trying to get industry to convert its operations to more sustainable methods while still earning a profit.

Chapter 5 discusses electronic products because this is an area that offers potential for energy savings in homes, schools, and businesses. The chapter describes innovations in small appliances, indoor lighting, and heat storage. It also discusses new technology using automatic systems to regulate indoor energy use. Finally, chapter 5 introduces *nanotechnology* and how this science of using ultrasmall materials might contribute to new energy systems in the future.

The next chapter describes new methods in designing landscapes that help buildings conserve energy but also minimize disturbances to nature. This chapter on ecological landscaping discusses the landscaping process, architecture, schemes for planting vegetation, and ideas for working with soil, water, and unique localized climates. Chapter 6 also looks at methods in rainwater conservation and new surface materials for driveways and walkways that conserve water. Finally, the chapter describes the specialization of landscape design, a profession that combines art with scientific training.

Chapter 7 covers new wastewater treatment processes that conserve water and energy in order to contribute to sustainability. It covers simple technologies and more advanced technologies in treating wastewater. The chapter also describes the usefulness of waste digestion by *microbes,* because, in addition to waste decomposition, microbial actions produce heat, gases, and solids that each can serve as *renewable energy* sources. The future of sustainable wastewater treatment is discussed, as well as the challenge of building this technology in the future.

Environmental engineering represents a discipline that will be required for almost all future technologies in energy conservation. This book shows how some of these engineering projects turn out to be quite complex endeavors, yet a good number of future projects will draw on the simplicity of natural systems by following nature's theme of less is more.

NEW DIRECTIONS IN CIVIL ENGINEERING

Environmental engineering is a branch of civil engineering that focuses on solving environmental problems. Civil engineering encompasses the design and building of structures and dates to ancient civilizations. The Egyptian pyramids, the Great Wall of China, and the Roman Coliseum are all accomplishments in civil engineering. Today's civil engineers develop roads, bridges, tunnels, buildings, manufacturing and power plants, airports, harbors, rail systems, oil pipelines, and water distribution and wastewater collection systems. To carry out these tasks, civil engineers receive training in computers, modeling and simulation, mathematics, physics, chemistry, geology, geography, and biology.

The land and how it moves determine the durability of a structure. Buildings constructed on unstable land or in flood zones would not have a safe, long-term future were it not for civil engineers, who either take these risks into account when developing their plans or decide simply not to build on such a risky site. Environmental engineers consider the surroundings too, but their planning focuses on the effect human actions have on the environment. Therefore, environmental engineers combine civil engineering with environmental science. Environmental engineering works to improve *ecosystem* health, manage pollution, and conserve natural resources. In truth, an increasing percentage of today's civil engineers emphasize the environment. Some engineering accomplishments of the past have been impressive edifices intended to last for many years, but these same structures gave little benefit to the environment. Universities that teach civil engineering now stress the need to work with and for the environment whenever possible.

Environmental engineers create structures to achieve either one of two goals: to help the environment or to minimize a structure's impact on the environment. The following traditional specialties in environmental engineering now take these goals into account: water treatment facilities; drinking water supply systems; wastewater and sewage collection systems; wastewater treatment plants; and nuclear power plants. Sanitation engineers play a role in all of these aspects of environmental engineering, except nuclear power, by targeting safe, efficient, and low-cost/low-energy water supply and wastewater treatment.

Recent advances in environmental engineering have created new opportunities in this field: *waste-to-energy* plants, *sustainable* housing and office buildings, fuel-efficient transportation systems, nuclear waste storage facilities, and city centers designed for minimizing fuel and energy use. Waste-to-energy operations involve processes such as incineration that create usable energy as they destroy wastes. Sustainable building is any type of construction that strives to conserve natural resources, both in the type of building materials used and in the final structure's operation. In the area of transportation, environmental engineers plan for roads and overpasses that reduce total driving mileage, road surfaces that reduce fuel waste, and transit systems that lessen the need for cars.

Environmental engineers must have knowledge of how landmasses move, the behavior of surface and groundwater, soil characteristics, and erosion. Environmental engineers cannot develop safe structures if they do not also consider the effects of natural events such as flooding, freezing, hurricanes, and seismic activity from earthquakes or volcanoes.

This chapter describes the growing discipline of environmental engineering by first covering its history and then describing important emerging areas. This chapter also describes new concepts in zero energy architecture and biomimicry. It provides a look at the main aspects of ecological design and discusses why environmental engineers increasingly turn to the natural world as a model for new designs and materials. The chapter then concludes with the techniques used in civil and environmental engineering today and technologies for the future.

HISTORY OF ENVIRONMENTAL ENGINEERING

Environmental engineering began with the first human settlements when people dug trenches to carry wastes, wells to draw drinking water, and cool underground pits to store food. Sewer systems may have been the earliest of all engineering projects in history. Such systems have been unearthed by archaeologists in Scotland, Mesopotamia (now Iraq), and Pakistan, all dating from 3000 to 2000 B.C.E. Later civilizations in Egypt, Palestine, Greece, and China built pipelines for hot and cold water, drainage systems, and even toilets. The Romans from 800 B.C.E. to 300 C.E. created water distribution systems so sophisticated that they provided a blueprint for modern systems.

Despite the Romans' stellar reputation in sanitation, Rome's citizens threw their share of waste into open ditches even as Roman engineers worked on new innovations in water and waste transport. Roman engineers successfully devised a way to reuse wastewater from public bathhouses to serve as flush water for toilets. The flushed water then proceeded to a sewer system.

The Romans also built an aqueduct system to carry freshwater 20 to 30 miles (32–48 km) from its source to Rome and other large cities. Over a period of 500 years, the empire's engineers built 11 separate aqueducts. Incoming water went to enormous cisterns situated at the highest points in a city, and pipes—they used lead pipes, now known to be a health hazard—distributed the clean water to public bathhouses, residences, and city fountains. Parts of the ancient Roman aqueducts remain today, and modern water distribution systems follow the general layout used in Rome centuries ago for supplying clean water to city residents.

After the Roman Empire declined about 2,000 years ago, no other society seemed as interested in basic sanitation. In fact, major feats of civil engineering ceased for the next 1,500 years. The great European cities that began expanding in the post–Roman Empire era ignored good hygiene and waste control. People turned away from science and engineering in the centuries following the Roman Empire due to an emergence of new philosophies. The Romans combined many of their theories with a desire to please the deities, but in the Middle Ages a notably nonscientific philosophy enveloped society in a way the Romans could never have imagined. Medieval astrology challenged any discoveries made in astronomy, and magic sometimes took precedence over medicine. The Christian era in Europe renounced many types of science and took the startling step of removing many scientific books from libraries and closing the libraries themselves!

Europe paid for its lack of attention to the scientific basis behind infection and cleanliness throughout the Middle Ages with a series of devastating plague epidemics that arose in the unsanitary conditions. Not until the 1830s in Paris did civil engineering take an important step forward, when a series of cholera outbreaks spurred officials to call for better sanitation, including effective sewers. Up to that point, Parisians blithely shunted wastes into cesspools that dated from the Middle Ages. In the face of more cholera outbreaks, engineers began laying pipe to better manage water flow and keep drinking water safe and separate from wastes. From the 1840s to the 1890s, Paris constructed an underground sewer system that became an exemplar for waste management throughout Europe.

Civil engineering projects had also been growing in the United States since George Washington's presidency. Washington has been credited with giving birth to the engineering profession due to his lifetime interests in land surveying and building design. Because the Industrial Revolution had not occurred in Washington's time, *engineer* was not yet part of the language. However, Washington applied engineering principles that are still used today. Washington planned and designed his estate in Mount Vernon, Virginia, in addition to barns, houses, and even farming equipment and canals, all to emphasize efficiency.

Environmental engineering developed further during the 1800s in England and the United States. Engineers who had participated in building the U.S. railroads turned their attention to the design of municipal sewer systems and drinking water distribution. President Theodore Roosevelt advocated conservation of the land and its natural resources throughout his administration (1901–09), but, other than water management, engineering had not yet begun to work in concert with environmental concerns. Environmental activism gained a voice in the 1950s and 1960s. When the author Rachel Carson decried the degradation of the environment due to *hazardous wastes* in her 1962 book *Silent Spring*, people began noticing an environment that had for centuries received

Aldo Leopold (1887–1948) was an internationally respected naturalist who wrote more than 350 articles on conservation and natural resource management. Leopold was one of many scientists who built the foundations of conservation in the United States and abroad. Scientists and writers like him paved the way for the environmental movement. *(Wisconsin Department of Tourism)*

poor care. Civil engineering also focused on environmental issues with increasing vigor.

Environmental engineering today goes beyond basic sciences such as biology or physics. The table on page 6 summarizes the main areas of expertise that contribute to environmental engineering.

The rapidly growing fields of green building and designing for sustainability have called on the skills of environmental engineers. All 50 states, the District of Columbia, and the U.S. territories now require engineers that serve the public to be registered with the Accreditation Board

DISCIPLINES USED IN ENVIRONMENTAL ENGINEERING

DISCIPLINE	ROLE IN ENVIRONMENTAL ENGINEERING
agricultural engineering	new equipment that reduces fuel use, soil erosion, and habitat destruction
architecture and design	construction of buildings that reuse natural resources and draw zero energy from a community's electricity supply
biology	the needs of plant and animal life in the natural environment
chemistry	the properties of synthetic chemicals that replace nonrenewable resources
chemical engineering	new chemical processes that require less energy than traditional chemical reactions and produce less hazardous waste
climatology	understanding how local climate will affect new structures
ecology and environmental science	air and noise pollution, waste management
geology	the effect of land structure and movement on new structures
hydrology	water systems that use minimal energy to supply households and are part of a reuse system
landscaping	building structures to work in tandem with the land's natural shape
materials science	selection of new building materials to replace rare or nonrenewable materials
mechanical engineering	new equipment that supports a building's functions with minimal use of nonrenewable energy sources

Discipline	Role in Environmental Engineering
physics	using gravity, force, temperature or other natural physical characteristics to replace energy-demanding mechanized systems
public health	devising waste and wastewater systems that are energy efficient without bringing health risks into a community
sanitation	proper management of waste flows and clean water and air to eliminate disease
soil science	understanding the characteristics of local soils to design efficient natural waste degradation systems and landscaping

for Engineering and Technology (ABET). The board requires newly certified engineers to hold a specialty in one or more of the following areas: air pollution, hazardous waste, *industrial hygiene, radiation* protection, solid waste, or water supply/wastewater engineering. Many engineers in the water or wastewater industries also practice specialties such as storm water management. Overall, the environmental engineering profession involves three types of assurance that an engineer has the expertise needed to design structures that are safe for people and for the environment, as follows:

1. ABET certification—an engineer possesses education, experience, and licensure in general environmental engineering or a specialty within environmental engineering
2. registration—an engineer has passed a competence examination and possesses training and education to be listed on a government or nongovernment agency roster of registered engineers
3. licensure—an engineer has been granted by a government agency the right to perform work in environmental engineering that directly affects the health and welfare of the public

BALANCING RESOURCES AND WASTES

Green communities refer to places in which residents conserve resources and energy while minimizing waste. Environmental engineers must understand the needs of people as well as ecosystems in order to develop these types of communities. The broadest definition of environmental engineering therefore involves the design of structures that accomplish two objectives: use natural resources in a way that conserves them and manage waste so that it cannot harm the environment. To meet these objectives, engineers design *sustainable loops* within structures so that resources are reused and wastes recycled. These loops work in homes, office buildings, factories, or towns.

Sustainable Loops

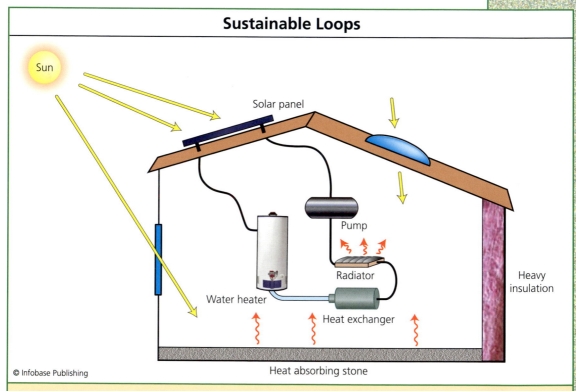

© Infobase Publishing

Sustainable loops make maximum use of energy in a building. The sustainable loop shown here circulates solar energy as heat to do work. Water is used to carry heat into the building's interior, and then cooled water returns to the solar panels to be reheated.

The terms *sustainability* and *green* have been used, and perhaps over-used, to a point where they can lose their meaning. Nicholas Low, Brendan Gleason, Ray Green, and Darko Radović explained in their 2005 book *The Green City,* "Sustainability does mean something both important and new. It speaks of the greatest change in human thought and behavior for 3,000 years. It is fundamentally about the global environment. It is also about cities." Specifically, sustainability has two major objectives that must both be met in order for it to be considered successful. First, sustainability must use the best methods at local and national levels for conserving all natural resources. These resources are water, soil, forest products, minerals, oil, clean air, and *biodiversity,* which is the variety of plants and animals found in a specific place. Conserving natural resources enables industries to reduce the raw materials harvested from forests or oceans or any other source, and this in turn preserves species habitat, which is the best means of conserving biodiversity. Second, sustainability must conserve natural resources in a manner that allows people to have a comfortable lifestyle, that is, adequate shelter, food, clean water, and health. Environmental engineers plan their projects so that they can meet both of these objectives.

ECOLOGICAL DESIGN

Ecological design, also called eco-design, enables environmental engineers to conserve nature while serving people. Ecologically designed structures follow these four principles:

1. They allow nature to be visible.
2. Their design works with nature rather than trying to reconstruct nature.
3. Design planning grows out of the characteristics of a location.
4. Designs conform to *ecological accounting.*

The principles of ecological accounting work similarly to a balance sheet in business accounting. Businesses develop balance sheets to assure that all of their expenses, or cash outflows, can be covered by the business's assets and cash inflow. Ecological accounting also will reveal whether a community's consumption and waste production can be covered by technologies that reuse, recycle, or reclaim materials. In environmental science, *natural capital* is equivalent to financial assets; resource depletion corresponds to financial expenses. An example of a financial balance sheet and an example of an ecological balance sheet are shown in the following table.

On the ecological balance sheet shown below, natural capital represents any natural resources that the planet holds and that are available for human use. Of course, unrestrained use of natural capital will quickly deplete resources so that these resources cannot be available for future

CREATING ECOLOGICAL BALANCE SHEETS

FINANCIAL BALANCE SHEET

Assets		Expenses	
cash	$3,000	utilities	$4,000
equipment	$10,000	rent	$12,000
vehicle	$4,500	bills due	$1,500
total assets	**$17,500**	**total expenses**	**$17,500**

ECOLOGICAL BALANCE SHEET

Natural Capital		Resource Depletion	
fuel	50,000 units	heating	40,000 units
water	100,000 units	electricity use	40,000 units
habitat	500,000 units	fuel consumption	640,000 units
forests	75,000 units	deforestation	5,000 units
total natural capital	**725,000 units**	**total resource depletion**	**725,000 units**

Passive solar homes use sunlight for natural lighting and heat. This house in Pennsylvania uses horizontal light shelves to reflect light into the rooms' interiors. The solarium that extends beyond the outer wall receives sunlight and reflects it. This serves the purpose of lighting the interior and providing heat. *(Cusano Environmental Education Center, John Heinz National Wildlife Refuge at Tinicum)*

generations. In order to make up the deficit, society must rebalance the accounting sheets in the following ways: (1) conserve natural resources; (2) produce new types of resources; or (3) combine conservation with new resource technology. The table on page 12 provides an example of balancing ecological accounts or, put another way, living sustainably. In all cases, the existing natural capital never changes because these values are set; the planet holds only a finite amount of natural resources that are gone forever once people have used them up.

In the example shown in the table on page 12, unrestrained resource depletion is solved by combining conservation with new technologies that provide alternative resources. Environmental engineering takes part in all of these options by making sustainable buildings, and it also develops new forms of resources, such as solar or wind energy, to help conserve nonrenewable coal and oil energy sources.

Environmental engineering helps balance ecological accounts in four specific ways. First, it uses construction materials that spare nonrenewable

BALANCING ECOLOGICAL ACCOUNTS

UNBALANCED ACCOUNTS

Natural Capital		Resource Depletion	
fuel	50,000 units	heating	60,000 units
water	100,000 units	electricity use	90,000 units
habitat	500,000 units	fuel consumption	900,000 units
forests	75,000 units	deforestation	8,500 units
total natural capital	**725,000 units**	**total resource depletion**	**1,058,500 units**

REBALANCED ACCOUNTS

Natural Capital		Resource Depletion	
fuel	50,000 units	heating	50,000 units
water	100,000 units	electricity use	65,000 units
habitat	500,000 units	fuel consumption	750,000 units
forests	75,000 units	deforestation	5,000 units
subtotal	**725,000 units**	**subtotal**	**870,000 units**
Technology			
biofuels	70,000 units		
solar energy	70,000 units		
alternative woods	5,000 units		
total natural and new capital	**870,000 units**	**total resource depletion**	**870,000 units**

or slow-renewable materials such as wood harvested from *old-growth forests.* Second, environmental engineering maximizes the use of *passive energy,* which is energy generated by nonmechanized means. Third, if a design must include active or mechanized energy, it employs it in its most efficient option, as, for example, replacing an older, energy-demanding refrigerator with an energy-efficient model. Fourth, environmental engineering balances production and consumption by working with the natural systems in the environment. For example, an ecologically designed house in northern Australia can make use of natural airflow for cooling and forgo air-conditioning. By contrast, a house on a Swiss mountain reduces wind exposure with a partially subterranean design that lets wind flow over it.

New structures develop in three phases: planning, design, and implementation. The best plans take into consideration all facets of the local environment. People have for generations built houses, barns, roads, and dams as if to battle nature rather than work with it. Expensive beachfront homes often slide into the waves during storms; roadways continually wash away in landslides. Better planning would allow the coast to change naturally in storms and keep homes secure by building them farther inland. Roadways that require continuous repair and support should probably not lie on unstable land in the first place. Designing with nature in mind also follows a ready-made blueprint: nature itself. The sidebar on page 16 "Case Study: How Do Prairie Dog Tunnels Work?" shows how engineers can learn from structures that are already part of nature.

Ecological design brings a new philosophy to almost every aspect of conventional design that has been used in cities and towns since the Industrial Revolution. The table on page 14 describes the differences between conventional design and ecological design.

A person who has lived a lifetime in a metropolis such as New York City might find it hard to understand the idea of ecological design because this person has grown accustomed to conventional design. Ecological designs follow few of the preconceived notions of an office or a house; they bring nature back into places that have been missing nature for generations. The Yale University professor Stephen Kellert explained in a 2004 interview that one of ecodesign's challenges is to incorporate nature in all places, not just in a garden outside a massive skyscraper. "One of the difficulties of the urban environment is that especially the large urban mega-cities around the world tend to rely ever increasingly on vertical structures," he said.

COMPARISON OF CONVENTIONAL DESIGN WITH ECOLOGICAL DESIGN

CHARACTERISTIC	CONVENTIONAL DESIGN	ECOLOGICAL DESIGN
ecological accounting	complies with environmental impact reports	tries to minimize all environmental impacts from materials choice to local ecosystems
energy source	consumes nonrenewable resources; relies on fossil fuels	sustainable loops in which wastes from one process feed another process to conserve nonrenewable resources; emphasis on renewable materials
toxic substances	commonly used	reduced or eliminated
design criteria	economics, custom, and convenience	prioritizes human and ecosystem health
sensitivity to ecology	standard designs applied over large, diverse regions regardless of culture or place	designs integrate with local climate, topography, soils, vegetation, local materials, and local culture
role of nature	designs are imposed on nature to control natural systems	designs work with nature and use natural structure whenever possible

Note: Adapted from Sim Van der Ryn and Stuart Cowan, *Ecological Design* (Washington, D.C.: Island Press, 1996).

"Most of our evolutionary experience of the natural world has been on a horizontal plane, so how do you build into a sixty-five-story office building an experience of nature that is any more than superficial or vicarious?" Ecological design intends to answer that question.

New ecologically designed structures, especially office buildings and housing, must overcome two challenges: the process of mimicking natural designs and the challenge of overcoming people's attitudes. Kellert has developed a classification system, shown in the following table, that describes nine attitudes toward nature that any person in industrialized society likely holds. The categories are listed in order from the most nature-aware attitude to the least nature-aware attitude. When perusing these categories, it becomes clear that many people will probably take a long time to accept ecological design.

Negativistic and dominionistic views of nature lead to conventional designs; greater appreciation of nature favors ecological designs. Sometimes ecological design means throwing out all previous notions of engineering to study the ways in which nature creates simple solutions to complex problems. The University of California professor Robert Full described to the *San Francisco Chronicle* magazine in 2008 his team's experience on

HUMAN ATTITUDES TOWARD NATURE	
CLASSIFICATION OR ATTITUDE	**VALUE SYSTEM**
moralistic	ethical and spiritual relation to nature
humanistic	emotional bonding with nature
aesthetic	physical attraction and appeal of nature
symbolic	nature as a source of language and imagination
naturalistic	exploration and discovery of nature
scientific	knowledge and understanding of nature
utilitarian	nature as a source of material and physical reward
dominionistic	mastery and control of nature
negativistic	fear of and aversion to nature

one difficult project: "They wanted to make a robot that maneuvered in the surf zone. This is really hard! It's a terrible environment to maneuver in, but of course crabs can do that very well." Full's colleagues had invented a complicated robot that needed at least eight motors on each leg to simulate a crab's joints. The entire blueprint became so unwieldy that the engineers decided to return to studying crabs to discover how these animals made

CASE STUDY: HOW DO PRAIRIE DOG TUNNELS WORK?

The ecological design expert Sim Van der Ryn once pointed out that "in a sense, evolution is nature's ongoing design process." Designers and architects had for decades been slow to understand this concept, especially in the industrialized world. Agrarian societies, by contrast, built structures more in keeping with the environment. In tropical Asia and Africa huts often stand on stilts to let breezes flow through and avoid moist ground, while desert dwellings consist of thick mud walls that cool interiors. It should come as no surprise that wildlife architecture, such as nests and dens, possesses the same type of melding with nature. Rodents called black-tailed prairie dogs that live on dry prairies and hot climate grasslands use a similar ingenious design with nature.

Black-tailed prairie dogs form colonies or towns of a dozen to a few dozen individuals that cooperate in hunting, as sentries, and for building burrows. Each burrow contains a complex of tunnels and one or more mounds. The tunnels and mounds shelter adults and pups, resist predators, and regulate temperature and ventilation. Prairie dog burrows may spread out to 100 acres (0.4 km²), but, because the animal's habitat has been dwindling in recent years, tunnels usually extend only 15 feet (4.6 m). Large or small, prairie dog colonies use the local environment to their advantage.

Prairie dog burrow entrances consist of dome mounds and crater mounds. A prairie dog builds a dome mound by digging upward from an underground tunnel until it breaks through the surface. The prairie dog spreads the excavated soil in a layer around the entrance and then pushes it into a dome that surrounds and sometimes covers the entrance. Meanwhile, other prairie dogs build crater mounds as secondary entrances and emergency escapes. The crater begins as a ring of uprooted vegetation, humus (organic soil), and mineral soil. The prairie dog waits for the next rain to moisten the soil and then compacts it, adding bits of plants to strengthen the mound. Such volcano-shaped crater mounds can tower up to two feet (0.6 m).

Prairie dogs situate the two different mounds at opposite ends of a burrow. When an aboveground breeze blows, air enters through the lower mound and leaves through the higher mound to cause unidirectional airflow through the tunnels, not only cooling the burrow but ventilating

locomotion so simple. "When we did that we found that all these different joints, many of them actually work together as one—synergistically, and so we could collapse those into a single leg." The *Chronicle* writer Mark Simborg said of Full's new approach, "They constructed a basic leg with two motors, and sure enough, it worked." The mantra less is more seemed to supply the answer.

it. But what happens inside very large burrows with intricate arrangements of thoroughfares and dead spaces? The dual mound construction creates air exchange throughout the tunnels due to a feature called a *velocity gradient,* which can occur in flowing air or water. A wind velocity gradient forms as wind blows over an open area. The air passing near the ground slows due to friction with the Earth, while air blowing at higher altitudes moves faster. These sheets of air supply the energy that pulls air into one burrow opening and out the other. (Because high-altitude winds move faster than lower breezes, they contain more energy. Therefore, the different-speed winds set up an energy gradient that corresponds to the velocity gradient.) The underground airflow moves at the same speed as the air blowing directly above Earth's surface, draws stagnant air from dead-end tunnels, and pulls in fresh air.

A prairie dog burrow with the air circulation system described here represents a zero energy design, which means that cooling, ventilation, and other functions occur without any need for energy input. The prairie dogs expend energy to build tunnels and mounds, but once those jobs are done, Mother Nature furnishes all the energy to run the circulation system. Prairie dogs even fine-tune their burrows by adjusting the shape of an exit mound's rim, depending on the local topography, wind exposure, humidity, and temperature. Sharp-rimmed mounds allow the most efficient airflow, so animals need these mounds for habitats with weak breezes or poor wind velocity gradients. In better wind conditions, the animals shape the mound into a blunt-style rim. Prairie dogs have therefore evolved into clever architects and likely seek places to live where the breezes are favorable for new burrows.

Humans can seldom achieve the prairie dog's skill in combining *form and function*, which means that the shape and size of a structure connects with how it works. Ecological designers can use animal dens, caves, or burrows to learn how structures work with or against local weather patterns and topography. Nature surely provides the framework for creating low-energy or zero-energy heating, cooling, ventilation, lighting, and water flow systems for new ecological buildings.

ZERO ENERGY ARCHITECTURE

Buildings account for 40 to 50 percent of all energy consumption and, in the United States, buildings produce more *greenhouse gases* than cars. One goal of environmental engineering is to develop new houses, offices, or manufacturing plants that minimize the amount of energy they use. A zero energy structure takes this goal a step further by not only minimizing energy use but by using no net energy at all in its operation. A zero energy building produces all the energy it needs to operate and requires no additional energy sources. A new discipline called zero energy architecture seeks to create zero energy homes, offices, museums, nature centers, and classrooms. When buildings have no need to draw energy from a utility, such as an electric company, the buildings are said to be zeroing out energy consumption or running *off the grid*. Put simply, zero energy architecture creates buildings that add up to zero on their ecological balance sheet.

Zero energy houses differ widely in style because they conform to local geography. Regardless of location, zero energy buildings have many of the following features in common:

- self-sufficient energy production
- emphasis on passive energy systems
- strategically placed shade trees for cooling
- added insulation from ivy and other plants surrounding the house
- south-facing windows to capture sunlight and heat
- skylights for natural lighting
- cross-ventilation from open windows and skylights

The Croatian-born architect Zoka Zola has built a zero energy house three miles (4.8 km) from the downtown Loop area in Chicago, Illinois. The house, completed in 2002, uses all of the features in the preceding list with two additions: (1) on-site solar, *geothermal,* and wind energy production, and (2) a green roof covered in growing foliage that absorbs rainwater and provides insulation. Said Zola in 2007 in an online video tour of the house, "The house is designed for two people to be a zero energy house, meaning zero fossil fuel consumption while the house is running." Similar zero energy houses have been built since then, but they remain relatively rare in U.S. communities.

An intangible benefit comes from ecological design and zero energy architecture: a new appreciation of nature for people who occupy eco-design buildings. People living in ecologically designed buildings begin to understand the benefits of natural lighting, ventilation, and water flow. In time, eco-design may lead people to change their behavior regarding conservation.

Zero energy concepts have branched out from single homes to entire communities, such as London, England's Beddington Zero-Energy Development (BedZED). Completed in 2000, BedZED contains more than 80 homes and businesses that form a carbon neutral community, meaning its operations put no net carbon into the atmosphere—the community absorbs as much or more carbon than it emits. BedZED accomplishes this by using the following main features:

- structural materials that store heat in winter and release heat in hot summers
- all buildings enclosed in 12 inches (30 cm) insulation jacket
- south-facing terraces to maximize heat-gaining exposure to the Sun
- north-facing offices to minimize Sun's heat that might be wasted
- emphasis on roof gardens, solar energy, sunlight, and wastewater recycling
- emphasis on interiors made of sustainable, recycled, or reclaimed materials
- capture of heat generated by occupants' activities, such as cooking
- meters in each room to help residents monitor their electricity use

Zero energy architecture requires even more commitment to energy savings than less strict ecological design. The concepts of zero energy design can be useful, however, as a model for conservation and sustainability in ordinary communities.

BIOMIMICRY

As the ecological designer Sim Van der Ryn mentioned, evolution has been architecture's best teacher. If the Earth's history were to be compacted into one year, humanity's time on the planet would be five and one-half minutes. Evolution had plenty of time to develop the best designs for most every structure on Earth without the need for computers or project management teams. The new specialty in ecological design and architecture called biomimicry relies on using designs invented by Mother Nature for structures for human use.

Biomimicry copies nature in three ways: form and structure, processes, or ecosystem operation. Designers who copy natural forms and structures concentrate on the shapes, sizes, and unique features of natural things and model their designs accordingly. For example, an engineer may design a stronger type of corrugated wall modeled on the corrugation of incredibly strong mollusk shells. Designers who copy processes, by contrast, follow the methods in which nature processes its structures. For example, eco-designers might develop water distribution systems based on the system found in redwood trees or even the human body. Redwoods transport water 350 feet (107 m) from their roots to the topmost upper leaves, and they do this silently, powered only by sunlight. The heart pumps blood through the cardiovascular system, and yet, as the California eco-designer Jay Harman pointed out in the *San Francisco Chronicle* magazine in 2008, "Your cardiovascular system: 60,000 miles [96,561 km] long and no straight pipes, and it's far more efficient than anything humans have ever dreamt of." These feats prove to be a significant challenge for *eco-building* designers.

The job of designing structures that mimic entire natural ecosystems presents a challenge even more difficult than mimicking a single structure in nature. A few zero energy houses achieve a facsimile of ecosystems by relying on sunlight for lighting interiors, solar energy for power, and perhaps a small wetland for cleaning wastewaters. Building a more complex ecosystem facsimile has been much more difficult. In 1991 eight scientists entered an enclosed structure called Biosphere 2 in the Arizona desert. They intended to demonstrate how humans could live in a self-contained unit that would mimic all of the Earth's nutrient recycling systems. Despite meticulous planning, the residents of Biosphere 2 struggled with an oxygen-depleted atmosphere caused by an overgrowth of microbes that

The wings of *Morpho rhetenor* reflect blue light due to nano-structures in the wing, which contains no pigmentation. The reflected light makes the wings appear blue to the human eye. The School of Electronics and Computer Science at the University of Southampton, England, used this butterfly wing structure as a model for nanoscale photonic crystals. Photonic crystals transmit information as photons or rays of light. *(Nano Group, School of Electronics and Computer Science, University of Southampton)*

consumed oxygen and emitted carbon dioxide (CO_2). This was the most serious but not the only problem they faced. Eventually, scientists inside and outside Biosphere 2 ended the experiment, knowing they had not mastered the ability to recreate a natural ecosystem.

Engineers may need to replicate nature's form and function on a small scale before attempting to mimic whole ecosystems. A clamshell offers a good example of the differences that exist between nature and humans regarding form and function. Clamshells have evolved into the perfect size to protect the clam, allow the creature to feed and eliminate waste, and withstand strong forces from waves and tides. The clam wastes no extra energy building a shell bigger than it needs. By contrast, people often design oversized houses that waste heat and energy and lead to overconsumption of products simply to fill up the space. Nature builds structures

Ecological design endeavors to mimic natural processes in structures used by humans. Copying an entire natural ecosystem is very difficult, but landscapers have been successful in re-creating small wetland ecosystems. A wetland like the one pictured here can clean runoff, act as a rainwater collector and irrigation source, and provide water for wildlife. *(CanadianPond.ca)*

perfectly suited for their functions. Sometimes nature makes structures flexible rather than rigid to absorb strong forces—trees that bend with strong winds—and in other instances the structure possesses incredible rigidity and strength—mussel shells that withstand constant pounding by ocean surf. Engineers copy the methods used in nature to increase a structure's durability. For example, engineers design skyscrapers and bridges to sway a little in the face of storms as trees yield to winds. Environmental engineers, nevertheless, still have some distance to cover in order to build things as well as nature. The table on page 23 shows the main differences between the characteristics of human-made structures and natural structures.

Ecological design should start with the question, "How would this problem be solved in nature?" Architects and engineers create strength and durability by building with steel and concrete, but nature often takes

CHARACTERISTICS OF HUMAN INVENTIONS VERSUS NATURAL DESIGNS	
HUMAN INVENTION	**NATURAL DESIGN**
geometric	organic, flowing
straight lines	curved lines
rigid	flexible and fluid
large	small
motion depends mainly on wheels, gears	motion depends on limbs

an entirely different approach. The design scientist Jay Harman relayed his observations on nature's ability to combine fragility and strength. "As I was swimming along the [Australian] reef, waves would come and I'd grab hold of seaweed so that I wouldn't be pulled onto the reef, and the seaweeds would break off, because they're quite fragile. And yet, time and time again, even in the most violent storms, I noticed these seaweeds wouldn't break off even with these huge waves coming past, so what's happening? Well, all these seaweeds were changing their shape to let the force go past." Some inventions modeled on things in nature include Alexander Graham Bell's telephone design based on the structure of the middle ear and George de Mestral's Velcro, which he designed to mimic burrs that stuck to his dog's coat. The table on page 24 describes additional examples of adaptations from nature.

Nature tends to use simple materials such as silica, calcium carbonate, or keratin (the protein that gives hair its strength), which all require little energy to make. Environmental engineers study not only the composition of natural materials, but also the way nature makes these materials. Designers and engineers take slightly different approaches toward mimicking nature. Designers use biomimicry to imitate natural designs. Engineers, however, stress the field of *biomimetics,* which uses engineering principles to build biology-based structures. The chemical engineer

NATURAL STRUCTURES FOR USE IN BIOMIMICRY		
NATURAL STRUCTURE	**ATTRIBUTE**	**POSSIBLE ADAPTATION**
eggs	minimum material, maximum strength, retains heat, shape allows return to original spot if it is pushed, all recyclable materials	building materials
abalone	extremely strong shell (see the sidebar on page 27 "Abalone Shell—Designed for Strength")	ceramics, building materials
cockroaches, centipedes	locomotion, climbing	robots for uneven surfaces
geckos	tiny adhesion hairs for holding onto vertical surfaces	adhesion of parts to one another without glues or mechanical connectors; self-cleaning materials
crabs	jointed legs for moving in surf zones	surf exploration robots
peacock feathers	iridescent pigments	mobile communication displays
butterfly wings	tiered structure creates absolute black	optics
bird primary feathers	shape changes for reduced *drag* and increased energy efficiency	airplanes, vehicles

NATURAL STRUCTURE	ATTRIBUTE	POSSIBLE ADAPTATION
bird bills	hard, strong, and sharp yet sensitive to stimuli	medical instruments
lotus leaves	microscopic bumps shed water and dirt	plastics, building surfaces, household surfaces
humpback whales	baleen	high-efficiency, self-cleaning filters
cochlear shell spirals	efficient use of space	stairways, water distribution systems, ventilation systems
spirals: whirlpools, tornadoes	transfer of large amounts of energy	low-energy pumps, propellers, and fans
boxfish	square, stable contour	fast, low-drag vehicles
sharkskin	tiny scales that allow water through while maintaining stability	fast, low-drag vehicles
thorny devil lizards	scales that wick (passively transport) water to mouth	water-capture and distribution technologies
termite mounds	temperature, humidity, and airflow regulation	zero energy buildings
glowworms	light production	cool, non–heat-wasting bulbs
deer antlers	fast-growing cells	studies on cancer cell growth or burn treatment
animal tendons	strength during bending, twisting, or pulling	materials for surgery, flexible surfaces, rubber, or plastics

Robert Cohen of the Massachusetts Institute of Technology explained to *National Geographic* in 2008 the value of biomimetics, "Looking at pretty structures in nature is not sufficient. What I want to know is, Can we actually transform these structures into an embodiment with true utility in the world?" This of course is the challenge of biomimicry and biomimetics. Harman warned, "It's like trying to understand the universe. In order to really understand the universe, we sort of have to be outside the universe and look at it from a distance. The best we can hope for is to achieve levels of understanding." The following sidebar "Abalone Shell—Designed for Strength" describes an instance in which engineers have studied the fine points of a natural structure to learn from it.

TECHNIQUES USED IN ENGINEERING AND DESIGN

Ecological design has grown out of a blend of engineering, architecture, and design. Engineering draws upon a variety of sciences to develop a structure to serve a specific purpose. Design encompasses scientific and artistic skills for planning what the structure will look like, and architecture is the building of the structure according to its design blueprint and engineering principles. These three professions have a responsibility to work together to create strong, lasting, and appealing structures for people.

Environmental engineering uses computer programs to plan construction projects, and designers and architects use a blend of computer programs and creative arts. All three professions employ the physical sciences, mathematics, chemistry, and the natural sciences. Appendix A summarizes the main subject areas covered in environmental engineering and design education. Environmental engineers then apply their training to specializations in engineering, described in Appendix B.

Design involves the use of technical drawing skills so that a designer can show a customer what a new structure will look like before it has been built. This craft of producing accurate and precise images of a planned structure is called *drafting,* and it consists of two types: manual drafting and computer-aided drafting. More than 200 years ago, new shops, barns, and houses began with a sketch of all the rooms, entrances, windows, and other pieces of the building. As society desired larger and more complex buildings, the drafting profession needed to produce drawings depicting exact measurements with all components in proportion to each other. To do this a drafter uses a drafting table and a set of drafting tools: T square straightedge, triangle, French curve, and compass. A T square

ABALONE SHELL—DESIGNED FOR STRENGTH

Nature creates many materials far stronger than any human-made materials, diamonds for example. Scientists who study biomimicry have found another paragon of strength in nature: the marine invertebrate abalone. Abalone produces a shell five times more resistant to breakage than synthetic ceramics, but the shell's composition is 3,000 times stronger than its main constituent, the mineral aragonite. Scientists have pondered how abalone manages this feat. "Its strength is very likely due to structure," said the University of Wisconsin physicist Pupa Gilbert to the LiveScience correspondent Corey Binns in 2007. Abalone shell is composed of a small amount of organic matter in a calcium carbonate material called nacre, which the mollusk builds into an irregular tiled plate. The abalone laminates the nacre plate within protein sheets; protein's positive charge enables it to bind to the negatively charged nacre. This layered shell protects the abalone against direct impacts but also provides another safety mechanism: The protein layers slide against one another to absorb energy from indirect blows. A cement mixer can roll over an abalone shell without causing even one crack.

In 2005 the mechanical engineer Kenneth Vecchio of the University of California, San Diego, succeeded in making an aluminum-titanium material modeled after abalone shell. Vecchio's formula has been used by the aerospace industry for providing steel-like strength to space vehicles while weighing much less. This same material may soon be used in military body armor. The University of Washington materials scientist Mehmet Sarikaya explained to the *New York Times* some of nature's mysteries: "Nobody knows all the details. Nature has designed an interface of organic and inorganic materials interpenetrating in ways that we cannot yet equal." Vecchio pointed out, "When you think about it, an abalone shell is just chalk." The abalone shell provides one of many examples in nature in which strength comes from simplicity.

Solar panels represent one of environmental engineering's most notable inventions. Solar panels, like these in Spain, have led the renewable energy movement by taking advantage of the massive amount of the Sun's energy that irradiates the Earth every day. Advances in solar panel technology will consist of higher efficiency solar collectors, thin solar films, and nanoscale solar chips that might be embedded in window glass. *(Fernando Tomás)*

is a T-shaped ruler, and a triangle is three rulers connected into a single triangular straightedge tool. A French curve also lies flat on the drafting table, but this tool contains a variety of curved shapes rather than straight edges. The drafter can trace an almost infinite variety of curved lines using a French curve. A compass is a two-legged instrument that enables a drafter to draw a perfect circle of a specific diameter; one compass leg holds a sharp point that anchors the circle's axis while the other leg holds a pencil for drawing a circle around the axis.

Manual drafting has given way to the use of computers to assist engineers, designers, and architects. Each profession relies on a computer technology called computer-aided design (CAD), which allows users to draw three-dimensional pictures on a computer screen. The drafter or designer views each CAD drawing on the screen and can rotate the image in any orientation to view the proposed structure's front, back, sides, top, and bottom, and from any angle. CAD uses a technology called *virtual reality,* which is a computer capability that creates artificial three-dimensional views. Put another way, a CAD user can view the interior or the exterior of a house from any perspective, just as if they were walking through the house or looking at it from outside. Virtual reality helps people imagine exactly what a structure will be, while it helps engineers evaluate proposed plans, detect structural flaws, and test the effects of weather, erosion, or seismic change.

CAD plays a twofold role in environmental engineering. First, CAD converts any hand sketch into a realistic drawing that obeys the laws of engineering. Second, CAD speeds design processes by allowing engineers to print a number of *prototype* designs, update any design to include the best features, and then reconstruct the new object into a three-dimensional image. The architect Michel Lewis said of CAD-maker Autodesk, "We can look at a design from all points of view and see the relationship between all our structures. Our digital model allows us to see every single detail, and to make changes really quickly. Without Autodesk, we would be working for another century and would not be able to achieve the degree of sustainability we wanted in this project [Playa Viva resort in Mexico]." Advances in CAD will continue to lead the way in environmental engineering.

CONCLUSION

The environmental engineering profession arose from civil engineering in order to accomplish two objectives: to create structures that serve people as intended, and to improve the environment or, at least, cause no harm to the environment. Environmental engineering encompasses many different areas of expertise such as drinking water and wastewater systems, ecological buildings, and energy-efficient vehicles. Today's engineers receive intensive training in subjects that describe the manners in which materials or built structures behave under varying forces from weather, climate, the land, or moisture, in addition to designing sustainable structures.

Sustainability ensures people do not consume more natural resources than the planet can replenish. Environmental engineers achieve sustainability by devising systems that reuse materials for new purposes, reclaim materials from wastes, and recycle waste materials to provide a benefit.

Ecological design achieves the same purpose by keeping track of ecological accounting, which simply is a way to assure that consumption does not outpace replenishment.

Ecological designers realize that structures in nature already contain attributes that maximize their usefulness and minimize the energy required to operate them. By mimicking nature's form and function, designers and engineers minimize the resources a new building will consume. This biomimicry holds a vast amount of potential for teaching engineers how to innovate by simplifying processes rather than complicating them. Indeed, simplification is a good starting point for building sustainability.

Designing with nature as the model requires a new way of thinking about how a house or a school or a factory should look. Many of the future's ecological designs may also look different from the familiar layouts of houses, offices, and buildings built years ago. In the past, traditional structures were built to be rigid, sturdy, and immovable, but the new designs will emphasize flexibility, delicateness, and movement. Most people will probably need some time to get used to these new ecological designs built on nature's blueprint. In this way environmental engineering is very much like other fields in environmental science: The science may turn out to be the easy part of the task compared with the job of convincing people to accept new ideas.

Designing Transit Systems

In the United States, vehicles emit carbon dioxide (CO_2) pollution in amounts second only to coal-burning power plants. The National Resources Defense Council (NRDC) has estimated that U.S. automobiles annually produce 1.5 billion tons (1.4 billion metric tons) of CO_2, which combines with other greenhouse gases to contribute to the world's current climate change.

New transit systems designed to support sustainability will be systems that reduce the waste of fossil fuels, decrease materials needed for building millions of miles of roads, and slow the buildup of atmospheric CO_2. Transportation experts who plan these systems must also consider the current layout of U.S. cities. Some metropolitan areas are *dispersed cities*; they grow outward and spread into formerly rural areas as their populations grow. Los Angeles, California, is an example of a dispersed city with a metropolitan area spreading more than 500 square miles (1,291 km²). Dispersed cities produce the familiar malady called *urban sprawl,* which leads to a greater dependence on cars. *Compact cities* by contrast grow mostly in an upward direction and travel involves shorter trips on public transit.

The New York City borough of Manhattan is a compact city covering only 23 square miles (59 km²). It grows upward in skyscrapers rather than outward, making this island the most densely populated place in the United States. Compact cities conserve land and so benefit the overall environment, but countries like the United States, China, Canada, and Australia with large open land areas tend to develop dispersed cities compared with other parts of the world. It may therefore come as no surprise that these three countries rank 1, 2, 3, and 4, respectively, in *carbon footprint,* meaning their activities depend on fossil fuels and their wastes produce large amounts of greenhouse gases. Dispersed cities affect carbon footprint in the following ways: (1) longer commutes burn excess fossil fuel; (2) excess fossil fuel combustion creates more greenhouse gases; (3) large numbers of commuters create traffic congestion, which wastes fuel and causes air pollution; and (4) a need for more roads and bridges, which use large amounts of building materials.

This chapter covers the important subject of transportation, including how certain characteristics of transportation damage the environment. The chapter explores new technologies that may change trends in transportation in industrialized nations. The main topics are the current state of transportation, personal vehicles, public transit, roads, and alternative vehicles. In addition, this chapter looks at the impact on the environment caused by rail systems, buses, trucking, airplanes,

This Los Angeles freeway has more traffic than it can handle, but Los Angeles is not alone in confronting traffic congestion. Most metropolitan areas now rate traffic congestion as one of their most serious problems. By developing advanced transit systems, cities can remove some cars from the road and reduce exhaust emissions. The most difficult obstacle to advanced transit systems is the American psyche, which has a strong connection to its cars.

and cargo ships. All categories of the transportation industry face the daunting challenge of making vehicles that will be less destructive to the environment than current models. Chapter 2 covers new approaches in how people and cargo get from place to place and the future of sustainable transportation.

TRANSPORT: CURRENT STATUS AND FUTURE NEEDS

Few people would dispute that personal vehicles cause environmental harm. The U.S. Department of Transportation's Bureau of Transportation Statistics (BTS) has estimated that at least 76 percent of commuters get to work each day in personal vehicles, and in some cities the percentage exceeds 90 percent. According to the Energy Information Administration (EIA) of the U.S. Department of Energy, 136 cars operate for every 1,000 people in the world, but in the United States the population has almost 800 cars per 1,000 people. Most of these vehicles carry a single driver; car pools account for only about 10 percent of U.S. daily commuters. The BTS statistics also show that in the United States people prefer cars to any other form of transportation, followed by trucks, motorcycles, long-distance buses, domestic air travel, local transit buses, freight transport rails, and commuter rails. The table on page 34 shows how this pattern of vehicle use translates to energy consumption.

Personal vehicles account for most travel. The heavy use of personal vehicles has the following three major effects on the environment:

1. Personal vehicles put a strain on nonrenewable fossil fuel reserves.
2. Vehicles demand an enormous amount of other resources to feed the automotive industry.
3. Vehicles hurt ecosystems and biodiversity by causing pollution and requiring extensive roadways, which destroy habitat.

Proponents of sustainability put considerable effort into getting people out of their cars and onto public transportation or alternative vehicles such as bicycles. If people must use cars, they should rely more on

U.S. ENERGY CONSUMPTION BY VEHICLE MODE	
MODE	ENERGY CONSUMPTION (TRILLION BTU)
trucks, all types	12,707
passengers cars and motorcycles	9,400
commercial airplanes	1,834
freight rail	581
general aviation	206
buses, long-distance	160
commuter transit	151
Source: Bureau of Transportation Statistics	

alternative fuel vehicles. California has led the way toward this goal by passing the 2008 California Complete Streets Act that requires communities to include multipurpose thoroughfares in all new roadway expansion plans. The bill's author, Assemblyman Mark Leno, remarked, "Streets aren't just for cars, they're for people, and with the Complete Streets Act local governments will plan for and build roadways that are safe and convenient for everyone—young or old, riding a bike or on foot, in a car or on a bus." In fact, new laws to build transit infrastructures might be the only way to convince people in the United States to leave their cars in the driveway. But it will not be easy: Cars provide convenience and a sense of freedom and independence that Americans find irresistible.

Sustainable transportation's success will probably depend on technologies that achieve two complementary outcomes: (1) fuel-efficient personal vehicles for drivers who cannot or will not give up their cars, and (2) attractive and efficient *mass transit* for everyone else. The Massachusetts Institute of Technology professor of mechanical engineering John B. Heywood has commented, "We've got to get out of the habit of thinking that we only need to focus on improving the technology—that we can invent our way out of this situation." In other words, new technologies that conserve natural resources will work only if accompanied by changes in behavior. "We've got to do everything we can think of," said Heywood, "including reducing the size of the task by real conservation." Sustainable transportation will need a four-pronged approach by targeting the following objectives: (1) reducing personal vehicle use; (2) developing alternative car technologies; (3) taking advantage of mass transit; and (4) developing compact cities.

URBAN TRANSPORTATION SYSTEMS

Efficient and attractive mass transit systems can greatly help in breaking the car habit. Some cities struggle with transportation systems that have noisy, polluting equipment that runs behind schedule and requires passengers to change vehicles to continue their routes. Clearly, commuters will not be eager to leave their cars for this experience. Other cities have built transit systems consisting of new vehicles that burn fossil fuel alternatives

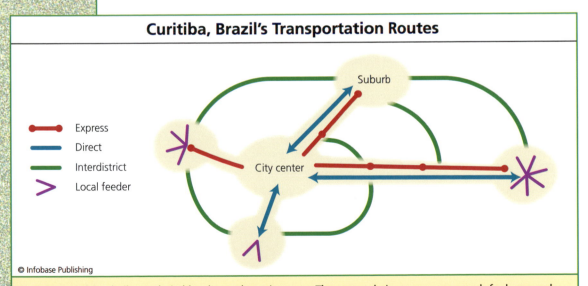

Curitiba, Brazil's Transportation Routes

- —•— Express
- —•— Direct
- —•— Interdistrict
- > Local feeder

© Infobase Publishing

Curitiba, Brazil, has built an admirable advanced transit system. The system designates separate roads for buses and long nonstop car commutes, point-to-point public transit routes, and local routes that speed the travel of people who do not need to enter the city center. The city now has what many consider the world's best bus system supported by well-planned highways and bus stations. Curitiba has also built extensive routes for bicycle commuting.

and that run on well-planned routes. Curitiba, Brazil, for example, emphasized transportation in 1969 when it embarked on a project to become an ecological city. Curitiba planned walkways, bicycle paths, bus routes, and car thoroughfares that complemented each other and took large numbers of commuters quickly into and out of the city center. The detail-oriented system in Curitiba even included extra-wide rail doors and platforms to allow more people to board and disembark at the same time.

Cities like Curitiba demonstrate that engineers can find viable ways to blend the use of personal vehicles, public vehicles, routes, and city layout. Few cities can reinvent themselves in such a total manner, but any success will likely come from a dependence on five interrelated features: (1) alternative fuel cars with limited use; (2) increased use of buses and light rails; (3) increased walkways and bicycle trails; (4) good roads; and (4) support activities such as bicycle parking structures, bicycle carriers on mass transit vehicles, and well-run, comfortable, and on-time mass transit.

Mass transit's main advantage over other modes of transportation resides in its ability to carry many passengers in a single vehicle. Despite this obvious benefit to the environment, mass transit must overcome many problems that hamper it in the United States and other countries, as summarized in the following table.

Future mass transit systems will need to fix the problems of noise, pollution, inconvenient routes, and poor performance, but new systems must also improve in other ways. People value certain intangible qualities such as convenience, safety, and cleanliness. Many individuals simply do not like mass transit's strict schedules, predetermined stops, and the feeling of riding in close quarters with hundreds of strangers. U.S. lifestyles also do not mesh well with mass transit schedules for the following reasons: (1) many people work long overtime hours and do not feel comfortable taking mass transit during nonpeak hours after dark; (2) mass transit does not lend itself to drop-offs and pickups at daycare facilities; and (3) mass transit is difficult to coordinate with extra trips linked with work commutes, such as shopping, medical appointments, and school schedules. Some rural areas have no access at all to mass transit even if riders would be willing to use buses and trains.

Perhaps the biggest hurdle to mass transit in the United States relates to habit. Some Americans see their cars as extensions of their identity. Developing countries, by contrast, have depended on mass transit as their main means of travel to work and recreation. The Institute for Trans-

ADVANTAGES AND DISADVANTAGES OF MASS TRANSIT SYSTEMS		
TRANSPORTATION MODE	**ADVANTAGES**	**DISADVANTAGES**
buses	can be rerouted to meet community needs; more flexible schedules than rail systems; less costly to develop than rail systems	noise and pollution; may operate at loss because of low fares to attract riders; vulnerable to traffic congestion; riders must travel by schedule
mass transit rails, subways, trams	energy efficient; reduced air pollution; reduces need for more roads and parking; reduces congestion; reduces traffic accident deaths	expensive to build and maintain; noise; riders must travel by schedule
rapid rails	quick connection between suburbs and city center; energy efficient	expensive to build and maintain; profitable only along heavily used routes; noise; limited scheduling options

portation and Development Policy (ITDP) based in New York City and Wohltorf, Germany, has developed the following primary focus areas for breaking the car habit:

1. developing high-quality, low-cost mass transit infrastructure
2. planning for and advocating cycling and walking
3. strengthening bicycle and other nonmechanized vehicle industries

Of the three ITDP goals listed here, high-quality and low-cost mass transit will be key to ushering people from their cars and onto buses and rails. Mass transit must outcompete cars by offering outstanding service.

This means improvements in comfort and safety, frequency of service, speed, and ease of transfer from one line or mode to another. Australia has focused on these things to make recent enhancements to its mass transit, and, as a result, transit systems have become profitable for large cities such as Sydney, Melbourne, and Brisbane. Mass transit improvements that have worked for Australia and that are now being adopted by other countries are the following:

- walk or cycle routes leading to park and ride transit stops
- mass transit vehicles that accommodate bicycles
- single ticketing system for an entire metropolitan area
- extensive suburban routes
- dedicated bus-only roads with overpasses at intersections
- buses with driver-activated priority at intersections
- feeder buses that link neighborhoods to express routes

Sydney, Australia's "T-way" transit system has used several of the improvements listed above to reduce some travel times by up to an hour. Yet even with these advantages, large countries like Australia and the United States do not maintain the mass transit ridership seen in densely populated European countries. In Brisbane, Australia, citizens take an average of 20 rail trips each year, while in Switzerland citizens take about 530 rail trips per year. Zurich, Switzerland, has developed a bus and tram system that is one of the best in the world, based mainly on the provision of separate roads for mass transit. In the early 1990s Zurich converted some of its streets to pedestrian and public cars only, removed some curbside parking to help traffic flow, raised tram tracks to avoid congested areas, and reduced the number of legal left turns—Swiss drivers drive on the right side of the road as in the United States. In addition, public transport vehicles control 90 percent of the traffic lights, which helps transit drivers make the route more efficient with less slowdowns.

No place on Earth is a mass transit utopia. Australian mass transit runs well, but still has its pitfalls; Zurich's system did not arise overnight and would likely be prohibitively expensive in the United States, with its much larger area. (The size of the New York metropolitan area alone is one-fourth the area of Switzerland.) American Public Transportation Association spokesperson William Millar told the *USA Today* reporter

European cities have developed the most efficient train stations in the world. Europeans travel by train more frequently than travelers in the United States. This station is Gare du Nord in Paris, France, which is the busiest rail station in Europe and one of the busiest in the world. *(iStockPhoto.com)*

Marisol Bello in 2008 that "only 5 percent of [U.S.] workers commute by public transit . . . and no more than 20 percent of households has easy access to buses or trains." If countries like the United States are to achieve sustainability in natural resource use, it will take a new commitment by people to mass transit, but also at a cheaper price tag. Joe Giuletti of the South Florida Regional Transportation Authority said in the same news article, "At a time that ridership is at an all-time high and people are desperate to use mass transit we are in a terrible spot." Mass transit remains a very tough test for the next generation of environmental engineers.

PERSONAL VEHICLES

Typical car advertisements illustrate the strong connection Americans have with their cars. Car ads often feature a vehicle cutting through mountain passes, speeding along coasts, or taking hairpin turns—never with another car in sight. Though these ads promote an association between driving and nature, cars actually threaten nature in many ways. Automobiles and trucks pollute the air with greenhouse gases and particles. Vehicles also create traffic congestion, which leads to additional pollution and prompts communities to build more roads. Then the domino effect increases urban sprawl.

Mass-produced road vehicles pioneered by Henry Ford changed lifestyles forever. Ford's innovation increased mobility and opened a new world of careers, learning, and communication, but it also soon produced congestion. Engineers worked diligently to design smooth roads for faster travel, plus traffic lights, bridges, and other deployments to keep cars moving. In the car's early history as now, drivers tussled with the thorny problem of getting around slow traffic. In the 1940s, engineers began planning divided highways in the United States to keep traffic moving; passing lanes allowed faster vehicles to overtake slower drivers. The divided highways, also called freeways, next included overpasses so that drivers could move through interchanges without stopping or even slowing. When urban sprawl accelerated in the late 1940s after World War II, city planners sat down with engineers to map out more roads based on patterns of trip origins and common destinations. New roads would bring, they thought, two rewards: income from vehicle and fuel taxes and a solution to congestion.

It is difficult to say whether new roads and rails encouraged urban sprawl or urban sprawl created the need for more roads and rails. In either case, additional road-building within the past few decades has not reduced congestion and, in many places, bad traffic has increased. The economist Robert Samuelson has been credited with the theory that "cars expand to fill the available concrete." People might choose to decrease their driving either because of its environmental impact or due to rising fuel costs (see the sidebar "Fuel Efficiency" on page 42). However, Colorado State representative Claire Levy pointed out to the *Denver Post* in 2008, "The average person can reduce their driving by only a small amount since it is impossible to get to work, school, church, or shopping centers without a lengthy drive." It seems as if Americans have become inextricably tied to their cars.

Economics influences driving patterns, as Representative Levy implied. Fuel costs often affect drivers' choices in personal vehicles, carpooling, or mass transit, but housing costs also play a part in transportation. Housing costs near the center of desirable cities such as San Francisco, New York, Denver, or San Diego force families to seek housing they can afford, houses that exist only in outlying areas. As a consequence, the number of commuters and the distances they travel to city-based jobs increase as housing costs in a metropolitan area increase. The *USA Today* writers Debbie Howlett and Paul Overberg explained in 2008, "To afford a house in a neighborhood with good schools, low crime and Saturday morning youth soccer, extreme commuters keep high-paying jobs in the big cities and buy houses well beyond the traditional metropolitan area. In California's Antelope Valley, across a mountain range from Los Angeles, commuters call it 'driving until you qualify.'"

The Zipcar car-sharing program at the University of North Carolina at Chapel Hill allows borrowers to reserve a car online on an hourly basis. An electronic reservation system remotely unlocks the vehicle at the time the reservation starts. Zipcar offers several on-campus pickup and drop-off stations. Several cities have experimented with similar car-sharing programs as well as bicycle-sharing to decrease traffic congestion and reduce overall emissions. *(GlobalExchange.org)*

(The *New Yorker*'s Rick Paumgarten once explained, "'Drive until you qualify' is a phrase that real estate agents use to describe a central tenet of the commuting life: You travel away from the workplace until you reach an exit where you can afford to buy [qualify] a house that meets your standards.") Mass transit has had a difficult time keeping up with these lengthening commutes.

Civil and environmental engineers understand that some drivers will not forsake their cars, and taking drivers out of cars may in fact hurt the global economy. The auto industry supports thousands of other businesses and also serves as a major recycler of metals. Many aspects of the world's economy—a mobile workforce, taxes, jobs, and tourism—depend on cars

FUEL EFFICIENCY

Fuel efficiency is the capacity to produce the greatest amount of work energy from the least amount of fuel input. In vehicles, this is called fuel economy. Fuel economy arises from the relationship between two types of energy: *potential energy* contained in a volume of fuel and *kinetic energy* associated with the vehicle's motion. When a car burns fuel, the car's potential energy decreases but its kinetic energy increases. Car designers try to design models that maximize this conversion so that most of the fuel's potential energy turns into kinetic energy and little excess energy disappears as heat.

Improved fuel efficiency will come from two components: the type of vehicles produced by car manufacturers and driving habits. Car engineers adjust the following characteristics to improve fuel efficiency: vehicle design, vehicle body's materials, and engine design. Vehicle designs and materials that increase the car's efficiency of movement help increase a car's miles per gallon (MPG), which is the main value that describes fuel efficiency. *Aerodynamic* design, lightweight materials, and an efficient combustion system all combine to increase MPG. Meanwhile, drivers must do their part by following the advice of the U.S. Department of Energy (DOE) regarding fuel-efficient driving. The DOE has estimated that certain driving styles can save on fuel use by the percentages shown in the following list:

1. observe speed limits, 7–23 percent
2. avoid rapid acceleration and braking, 5–33 percent

too much to ignore. Car-sharing programs offer a good compromise to the abundance of single-driver cars. In these programs, drivers register with a car-sharing enterprise in their city and reserve a car at a set location and then leave the car with its keys once they are finished driving it. The next car-sharer takes the vehicle from there. The automotive analyst Thilo Koslowski told the *Boston Globe* in 2007, "The next generation of drivers may have a little bit different view of how to meet basic transportation needs—they may not need to own a vehicle." For the present, car-sharing is not offered in every city so it has room to grow. But car-sharing's appeal lies in its potential to reduce the total number of cars on the road while allowing drivers the independence they enjoy.

3. avoid excessive idling

4. remove excess weight, 1–2 percent

5. use cruise control

6. use overdrive gears

7. keep engine tuned, 4 percent

8. replace dirty air filters, 10 percent

9. keep tires at recommended pressure, 3 percent

10. use vehicle's recommended grade of motor oil, 1–2 percent

11. use trunk for storing items rather than roof rack, 1–2 percent

12. choose drive times to avoid peak congestion

13. combine trips

14. select vehicle based on fuel economy

15. buy a gas/alternative fuel hybrid car

All of the recommendations on this list cause few inconveniences for drivers and are easy adjustments to make. Drivers need only be willing to change their behavior in minor ways to help reduce harm to the environment and also save on their fuel expenses.

PEDESTRIANS AND PARKING

The cleanest and healthiest way to traverse neighborhoods and the downtown is as a pedestrian. Many European and U.S. cities have made safe walkways to help pedestrians get to their destinations, but other places have become notorious for their lack of amenities for walkers. As people stop walking and increase their use of cars, not only does the environment receive more air pollution but people's health also declines.

Reid Ewing's research team at Maryland's National Center for Smart Growth reported in 2003 in the *American Journal of Health Promotion* that U.S. urban sprawl contributed to obesity, high blood pressure, hypertension, heart disease, and diabetes—results that seemed to point to the lack of walking done by people in the study. Ewing explained in a 2006 interview published on ScienceWatch.com, "Using health data from the Centers for Disease Control and Prevention's Behavioral Risk Factor Surveillance System, this study showed that American adults living in sprawling counties walk less, weigh more, and are more likely to be obese, and are more likely to suffer from high blood pressure than otherwise comparable adults living in compact counties (after accounting for individual socioeconomic and behavioral differences). This was the first study to show a link between urban form and the U.S. obesity epidemic." Ewing's study highlighted the often overlooked fact that mode of travel affects health and even the country's healthcare costs.

European engineers have designed walking centers in several cities, including Copenhagen in Denmark, Cologne in Germany, Montpellier in France, and York in England. City planners can either design new car-free zones that admit only pedestrians and bicyclists or close off streets for periods of time to create temporary car-free zones. The sidebar on page 46 "Can Bicycles Make a Difference?" takes a closer look at bicycles' contribution to sustainable communities. Of course, roping off streets to exclude cars requires additional steps such as setting up access for business deliveries and providing parking for cars that reach the perimeter of the car-free zone. Redesigning cities to improve transportation is therefore a large task.

Parking availability has an important impact on the environment also mainly because it helps gets cars off the road quickly, allowing drivers to turn off their engines. Conversely, congestion caused by lack of parking spaces creates streets backed up with idling engines that emit exhaust. Multistory parking structures are better than large parking lots, which cause the following problems:

1. remove large amounts of land from nature
2. contribute to rain runoff and soil erosion
3. absorb oils, fuels, and road salts that then wash away with runoff
4. hold heat and contribute to warming cities

Environmental engineers design new ecologically friendly parking garages to reduce the harm that sprawling parking lots have caused to the environment. The table on page 46 describes innovations for new community parking garages.

In California, the Santa Monica Civic Center parking garage provides convenient parking with many of the attributes listed in the table on page 46, plus additional innovations. Santa Monica's garage is the nation's first Leadership in Energy and Environmental Design (LEED)–certified parking structure because of these innovations: a storm water drainage and treatment system for use in the sprinkler system; use of recycled construction materials; coatings and paints that produce a minimum of toxic vapor–producing substances; surface glazing that promotes heating and cooling efficiency and reflects light to reduce electricity use; rooftop solar panels that generate one-third of the garage's energy needs; and energy-efficient mechanical systems. The CNN reporter Desa Philadelphia noted, however, in 2008, "Those features weren't cheap: the $29 million price tag for the car park is about $10 million more than for comparable garages. Santa Monica estimates that its parking structure will be profitable within fifteen years [of its construction]." Even parking garages have a place in sustainable communities if planners manage the costs carefully.

Conveniently located parking structures help ease traffic congestion by reducing the time a driver spends searching for downtown parking. Garage builders often include energy-saving lighting, architecture that admits daylight, and recycled building materials. This Ecoplex parking garage in West Palm Beach, Florida, included several principles of green building in its construction and use. *(Tilt-up Concrete Association)*

INNOVATIONS IN PARKING GARAGE DESIGN

INNOVATION	SUSTAINABILITY VALUE
mixed-use	parking in same facility as offices, shopping, and residences reduces overall driving
special parking zones	encourages use of bicycles and motorcycles
car-sharing zones	cars shared by more than one driver
pedestrian access	open, safe, and inviting walkways encourage garage use
automated vehicle	speed payment and flow through the garage, eliminating idling
identification systems	lines at the pay station
outlets	encourages plug-in electric vehicles
new materials, passive lighting, solar energy	nonrenewable resource conservation

COMMUTER RAILS AND BUSES

Commuter rails and buses offer the advantage of carrying many commuters on the same vehicle on a single trip so that fewer personal cars crowd the roads. Commuter rails have evolved over the past 100 years to encompass a variety of styles seen in cities today. Some large cities contain more than one mode, and almost all large U.S. cities contain at minimum a bus system and a rail system. The table on page 49 explains the types of rail systems in use in the world today; all have in common the requirement of running solely on set routes determined by preexisting tracks.

Heavy rail refers to railroads that carry freight and passengers coast to coast. Some railway lines combine freight cars and passenger cars, but others such as Amtrak trains run mainly for long-distance travelers and commuters. Local commuter rails, such as New York's Long Island Railroad

Can Bicycles Make a Difference?

Bicycle commuting requires similar thoroughfares as those used by pedestrians. That is, bicycle routes must be safe, well-lighted, well-paved, inviting, and go where the traveler wants to go. The benefits of using a bicycle to replace cars in most commuter trips should be familiar to almost everyone: bicycling promotes health; bicycling reduces air and noise pollution; bicycles do not use fossil fuels; bicycling relieves traffic congestion, which burns more fuel; and bicycles do not require much space to park—about 20 bicycles can park in a single car parking space.

Bicycling Life, a Web site devoted to bicycles as an alternative mode of travel, has noted that bicycling gives added benefits beyond good health: bicycles are inexpensive for owners and for city infrastructure; high-quality bicycles include more advanced material and power technologies than cars; and in urban centers bicycles reach their destinations faster than cars, but bicycling has not become part of global industry. The last point refers to the wariness that many people feel toward giant corporations that have close ties to government. In 1955 General Motors chairman Charlie Wilson proclaimed, "What is good for General Motors is good for America." At the time, Wilson made a valid point because the country's economy and car-oriented culture brought new post–World War II conveniences and wealth. But the auto industry also grew into a huge enterprise that collected as many critics as it did supporters. The bicycle industry, by contrast, supports local businesses and has not become tied to political agendas.

A new bicycle culture in the United States may take different forms that will all help the environment. First, commuters can substitute bicycles for cars whenever possible. Bicycle travel may be point to point or it may be composed of partial trips in which a bicyclist rides to a subway or bus station and then travels part of the route on mass transit that also carries the bicycle. Second, bike-sharing works the same as car-sharing: More than one person makes use of a single bicycle in a day, either free or with a small rental fee. After a few hours, the rider returns the bicycle to the same drop-off/pickup location. Bike-sharing in Paris, France, outshines any other place in the world for efficiency. In 2008 the *Time* reporter

(continues)

(continued)

Kristina Dell wrote, "It's hard to walk more than two blocks [in Paris] without running into a bike rack, which helps explain why the program has already yielded a 5 percent drop in car traffic. Paris has also removed lots of parking spots to make way for bike stations." Lyon, France; Copenhagen, Denmark, and Barcelona, Spain, also run successful bike-sharing programs. By comparison, Washington, D.C.'s program has 120 bicycles and only 10 stations. A third bicycling option involves the use of electric bicycles, also called e-bikes, that run on power levels much lower than those consumed by cars—150 *watts* of electrical energy versus 15,000 watts for a car.

Bicycles certainly make a difference in preserving the environment, especially if people combine bicycling with other sustainable activities. To help bicycling grow and make a true impact on the environment, some of the following issues must be resolved: (1) safety for bicyclists in heavy vehicle traffic; (2) building more bikeways; (3) options in bad weather; (4) theft prevention; and (5) developing respect between drivers and bicyclists. As the *Time* reporter Dell pointed out, "With gas prices skyrocketing and carbon-footprint consciousness going mainstream, more and more cities are betting that Americans are finally ready to make biking part of their daily commute." Dell's prediction has already come true in many cities and towns where a small but dedicated subpopulation of commuters has made the bicycle their primary mode of travel.

(LIRR), cover shorter distances, and, in many instances, these local lines bear the majority of rail travel in a metropolitan area. For instance, the LIRR carries almost 300,000 commuters each weekday and is the busiest commuter railroad in North America.

The difference between heavy rail and light rail can be difficult to understand at times because many light rail systems serving cities and suburbs resemble long-distance commuter rails. The following two principles have helped clarify the difference between light and heavy rail, but even these principles have exceptions and may be thought of as merely rough rules of thumb. For instance, most subway riders board trains from a platform.

1. Light rail travels 10 miles per hour (16 km/h), up to 60 miles per hour (96 km/h), and heavy rail travels faster than 60 miles per hour.

2. "On heavy rail, you board the train from a platform. On light rail, you board the train from the ground." (engineer Harry H. Conover)

TYPES OF RAIL SYSTEMS		
RAIL SYSTEM	**DESCRIPTION**	**PRINCIPAL USE**
light rail		
trolley	single car, about 35 passengers	city centers
streetcar	single car, 60–75 passengers	city centers and neighborhoods
tram	single to multiple cars, less than 100 passengers	short point-to-point routes, such as hotel to shopping center
light rail	multiple car, up to 250 passengers	city centers and suburbs
subway	multiple car, more than 250 passengers	city centers, all or part of the route underground
heavy rail		
freight rail	freight (boxcar and flat) cars, tanker cars	commerce
commuter rail (Amtrak)	multiple car, more than 250 passengers	long-distance, region-to-region, high-speed travel
commuter rail (local)	multiple car, more than 250 passengers	long-distance, city-to-city, high-speed travel

Trolleys, streetcars, and trams differ from rails because they usually travel at low speeds, and they often receive electric power from lines above the car. These modes alleviate some downtown congestion, but they may also contribute to traffic congestion in certain places such as crossings.

Light rail covers longer distances from outlying areas to city centers and presents the best features for getting people out of their cars and onto mass transit. Light rails usually travel on tracks in which the electrical power comes from a dedicated power rail in the track, sometimes called the third rail. Traffic design engineers now develop light rail systems that share some of the characteristics of streetcars and trams, such as routes along mixed traffic streets, dedicated rights of way, exclusive corridors, or in the middle of major thoroughfares.

Light rails are easier to build than heavy rails and have lower operating costs. These advantages enable engineers and local governments to plan convenient routes to reach commuters yet avoid wetland and woodland environments. Light rail and subway systems hold great potential for conserving fuel in urban areas and reducing harm to the environment, especially by incorporating the following points:

1. different types of systems to expand commuter choices
2. convenient stations
3. convenient commuter-hour scheduling
4. on-time performance
5. minimize breakdowns
6. preservation of open space

Commercial bus lines strive for similar objectives as light rail systems, that is, moving large numbers of people swiftly on a single vehicle. Long-distance bus companies such as Greyhound run intercity bus lines, while transit buses (also called urban or city buses) serve neighboring towns or a single large city. Buses have two advantages over light rail: (1) buses often serve rural communities where no other mass transit system runs, and (2) bus routes can be changed according to a community's needs. Like light rail, buses carry as many as 100 passengers on a single vehicle (also called a motor coach, omnibus, or autobus), which spares roads and fuel, but they also congest urban traffic and produce emissions. Unfortunately, buses have a long history of running on diesel fuel or gasoline and so have contributed to greenhouse gas buildup.

Towns in the United States have made efforts to convert their public buses and school buses to cleaner technologies. Lynn, Massachusetts, for example, plans to retrofit by 2010 all its 5,500 school buses so that filters clean pollutants out of engine exhausts. Ed Coletta, spokesperson for the state's Department of Environmental Protection, said, "These buses are going to be used long into the future, and we want to make sure they're emitting as few gases as possible." Lynn joins hundreds of other communities that have tried to balance the advantages and disadvantages of buses with a hopeful future.

New clean technologies can apply to vehicles in addition to transit, school, and shuttle buses. The DOE's Energy Efficiency and Renewable Energy Program keeps track of various public transit fleets that have made similar inroads into pollution control and alternative fuels: delivery services, long-haul trucks, refuse haulers, taxis, rental cars, and police vehicles. All of these vehicles offer the advantage of belonging to fleets that return to a central base so that mechanics can assure the vehicles receive proper maintenance and fueling. The following table describes innovations that are rapidly emerging in public bus fleets.

FUEL EFFICIENCY INNOVATIONS IN BUSES	
INNOVATION	**DESCRIPTION**
alternative fuels	biodiesel, ethanol, propane, compressed natural gas, electricity
hybrid vehicles	conventional fuel engine combined with electric-powered engine
fuel cells	hydrogen-based power units
anti-idling technology	auxiliary power units for electricity, automatic engine on-off equipment
manufacturing	light- and medium-duty vehicles for low-passenger routes, lightweight components, and vehicle frames

Advanced transit represents a new type of bus or rail travel that uses more than one technology for the purpose of conserving energy and fuel. For example, a community that is trying to use energy and resources in a sustainable manner may develop an advanced transit system, also called light transit, that contains the following features:

1. a network of interconnected mass transit routes for buses, light rail, and heavy rail
2. dedicated commuter lanes
3. alternative fuel-powered fleets
4. electrical power hookup stations in addition to the fleet's base station
5. express routes to bypass local station stops
6. emphasis on speed throughout the network
7. accommodation for bicycles in all fleets
8. easy and fast line switching, including long platforms and wide doors

Using the features above, advanced transit can meet its goals of eliminating inefficient schedules and excess fuel consumption, all while providing an enjoyable experience for riders. To do this, future advanced transit systems will likely include vehicles that are not yet common in today's mass transit, such as trams, light-duty buses, dedicated bicycle and scooter routes, and extensive use of car- or bike-sharing in city centers.

AIR TRAVEL

Air travel fuel efficiency has improved in the last 30 years more than threefold, and since 2000 fuel conservation in the airlines has improved by almost 25 percent. The U.S. passenger fleet has made even better strides when measured in revenue passenger miles (RPM), which equals one fare-paying passenger carried one mile. The U.S. RPM per gallon of fuel has more than doubled since 1978. Passenger and freight airlines made this remarkable progress in fuel efficiency by using the following methods:

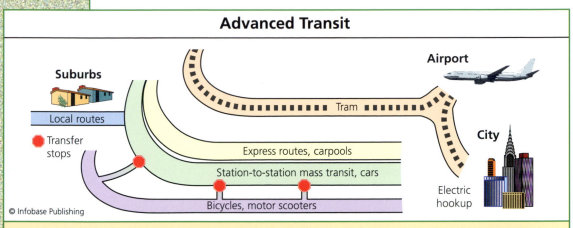

Advanced Transit

Suburbs

Airport

Local routes

Transfer stops

Tram

Express routes, carpools

Station-to-station mass transit, cars

City

Electric hookup

Bicycles, motor scooters

© Infobase Publishing

Engineers design advanced transit systems to meet the particular needs of a city and its surrounding communities. The goal of advanced transit is to speed travelers to their destinations with superior efficiency. Future advanced transit projects will include safe routes for pedestrians, bicycles, and motor scooters, and they will provide electricity for recharging electric-powered vehicles during the workday.

1. single-engine taxiing
2. selective engine shutdown on ground delays
3. better weight distribution
4. cruising longer at higher altitudes
5. shorter, steeper approaches to cruising altitude

Aircraft engineers may always struggle to simultaneously solve fuel efficiency and emissions because high-altitude flights give better fuel efficiency, but slower aircraft cruising at lower altitudes emit less CO_2. Airline industry engineers, nevertheless, have other opportunities that can complement each other for reaching better overall efficiency, such as the following:

1. improved navigation and weather forecasting systems to calculate most efficient routes
2. fuel-efficient engines and aircraft
3. *winglets* to reduce air drag and reduce fuel use 3–5 percent
4. new paints to reduce heat absorption, which requires extra energy for cooling

5. airport power hookups rather than use of onboard auxiliary power when at the gate

6. innovations in current kerosene-type jet fuel that comes from crude oil

7. aircraft innovations to reduce noise pollution

Redesigned hubs (a centralized heavy-traffic airport) and better flight scheduling should also reduce resource waste. For instance, a typical airport may run the following operations in a single day: (1) jet fuel arrives by pipeline or truck to a storage site; (2) airport hydrant system distributes fuel to the terminals; (3) hoses carry the fuel from the terminal hookup to individual planes; and (4) refueling trucks carry fuel to remote terminals. These systems work well enough, but newer designs will extract savings by relying less on trucks or by changing an airport's capacity to fuel more airplanes from a single fuel tank. Airports have already changed configurations to allow more planes to land and take off without ground delays, which saves fuel on the ground and in the air. The Air Transport Association of America has stated its view on fuel efficiency: "Beyond the numerous, diverse, successful measures that U.S. airlines have taken and continue to explore to conserve fuel, the single biggest advance in fuel conservation, and emissions reduction, will come from reform of the U.S. air traffic control system, which continues to rely on 1950s technology and procedures." Clearly, the airline industry has a broad horizon of opportunities for improving society's use of natural resources.

ROADS

Two aspects of road construction affect fuel efficiency: the road's surface and the road's length. Smooth roads reduce the power vehicles need to maintain their speeds by reducing surface-tire rolling resistance. Environmental engineers therefore design durable roads that can carry heavy loads without cracking or wearing, yet give vehicles a smooth driving surface. Asphalt and concrete meet these needs but they also affect the environment in negative ways. Asphalt comes from the crude oil refining process, so it consumes nonrenewable fossil fuels, and the production of concrete's main ingredient, cement, accounts for 7 to 8 percent of the world's carbon dioxide emissions. New road building in sustainable communities will therefore focus on three areas for improvement: (1) alternate road materials; (2) road planning; and (3) road use.

The road construction industry constantly seeks new road materials to hold down paving costs, and for that reason road builders already mix waste materials from other industries with asphalt or concrete. The following wastes in pulverized form work well in roads: crushed glass, tires, rubber, plastics, concrete wastes, and incinerator ash.

Road planning and design work hand-in-hand with alternative materials to reduce roads' carbon footprint. For example, gas-wasting designs such as long cloverleaf entrances and exits on freeways can be phased out and replaced with shorter but safe entry and exit lanes. Newer roads also improve water conservation with drainage ditches that catch runoff and channel the water to wastewater treatment plants. Future road materials may soon be used in which a porous surface allows rainwater to seep into the earth and thereby reduce the burden on treatment plants.

Road planning also involves study on the ways people use roads. Commuters today follow very different driving patterns than commuters of a decade or two ago. The U.S. Department of Transportation's Bureau of Transportation Statistics (BTS) has conducted a study that related a region's economy to driving patterns. The BTS found a recent rise in the proportion of daily "stretch commuters" or "extreme commuters," meaning they travel more than 50 miles (80 km) each way to work every day in trips that take 90 minutes or more. Stretch commuters make 19 out of 20 of these trips in personal vehicles rather than mass transit. Housing costs have often contributed to these lengthening commutes due to the "drive until you qualify" syndrome. The MSNBC reporter Allison Linn wrote of a familiar circumstance in 2008: "For Dollie Kinkead, the economic turmoil gripping the country translates into an 80-mile [129 km] drive each workday from a house she can't sell to a job she thinks she's lucky to have." Environmental engineering can help alleviate this bleak picture, but factors in society obviously play competing roles.

Long commutes may not be as fixable as finding new technologies in road building and planning. Austroads is an association of Australian and New Zealand road and traffic experts that takes a three-pronged approach to road-planning: (1) drivers' needs; (2) economics; and (3) the environment. This organization wants to foster sustainable roads, meaning roadway construction that reduces resource consumption, improves landscape and air quality, does not cause water or noise pollution, and

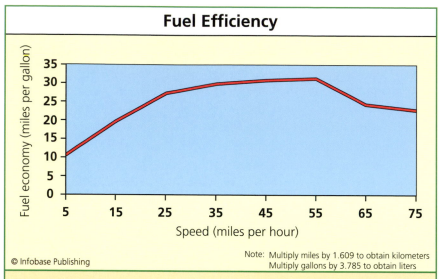

Fuel Efficiency

Note: Multiply miles by 1.609 to obtain kilometers
Multiply gallons by 3.785 to obtain liters

© Infobase Publishing

The typical combustion engine converts gasoline to motion efficiently at speeds higher than 35 miles per hour (56 km/hr) with steady driving to avoid fast accelerations. Fuel efficiency decreases in many cars at higher speeds.

protects habitats. Austroads has taken on all of these goals while providing resources for building quality thoroughfares for drivers and mass transit. Austroads's chairman Alan Tesch explained in the organization's latest annual report (2007–08), "A total of 41 projects were completed [in the past year] and 29 research reports and technical reports published. There were also 15 guides published . . . The guides are intended to be used by road authorities in Australia and New Zealand as a complete package of technical information that applies a consistent standard to road building . . ." Austroads also offers technical resources on road design, planning, maintenance, and operation.

In the United States, engineers have reexamined two road designs: cloverleaf interchanges and traffic circles. A cloverleaf contains four loops and one or two bridges called overpasses that enable drivers to switch from one roadway to another without stopping at a traffic light. These designs decrease time spent idling at lights, but they also lengthen driving distance. Traffic circles (called roundabouts in other countries) allow traffic to flow without stopping at interchanges between two or more roadways. Circles have offered an efficient traffic control device for many years until the 1980s when heavy traffic overwhelmed many circles. Densely populated New

Jersey, for instance, once depended on more than 100 traffic circles state-wide to ease traffic flow, but the state has gradually removed many of its busi-est circles that slow traffic rather than speed it up.

This chapter points out that even the best planning in innovative materials and road design will prob-ably not change human behavior. People like to drive, and drivers do not always travel in ways that are best for the environ-ment—they use excessive speed, abrupt starts, idling at drive-up windows, and

Environmental engineers today develop ways to decrease fuel waste while maintaining safe roads. This interchange in Fort Worth, Texas, in 1958, seemed like a well-designed answer to navigating the country's new interstate highway system. This design called a cloverleaf, however, requires extra driving and road-building materials so it may not be the best choice for helping the nation reduce its total fuel consumption. *(TexasFreeway.com)*

unnecessary trips. The strong connection between people and their cars is discussed in the following sidebar, "Case Study: The World's Growing Car Culture."

FREIGHT TRANSPORT

Railroads have been a major mode of cross-continent transport since the mid-1800s. In 1869 the Union Pacific and the Central Pacific railroads met at Promontory Summit in Utah to complete the United States' first trans-continental railroad. That single event launched a generation of rail cargo shipments and passenger travel. Today, the U.S. freight railroad indus-try operates on almost 145,000 miles (233,000 km) of tracks and carries about $37 billion of freight. Freight trains carry mostly coal (21 percent of their traffic), containers or truck trailers (14 percent), and chemicals (12 percent).

The Federal Railroad Administration (FRA) has cited four benefits to the environment from trains: (1) railroads are more fuel efficient than trucks, and fuel efficiency improves each year; (2) a locomotive emits about

CASE STUDY: THE WORLD'S GROWING CAR CULTURE

The social observer Ivan Illich wrote in 1974, "The model American male spends more than 1,500 hours per year on his car; driving or sitting in it, parking or searching for it; earning enough to pay for the vehicle, the tolls, the tires, the insurance or highway taxes." Things may have changed little since then other than American women joining men in the love for their cars! The situation is known as *car culture* in which a society's activities—economy and recreation—revolve around the car, but so too do people's sense of style, wealth, or identity.

The car culture is composed of the production of automobiles but also includes car-related products, fuel production, fuel consumption, and car travel for work, school, shopping, and vacation. Of the fuel consumption component of car culture, President George W. Bush declared in 2006, "Here we have a serious problem: America is addicted to oil, which is often imported from unstable parts of the world." Bush made a valid point; the United States consumes nearly one-fourth of the world's oil.

The car culture flourishes in the United States more than any other nation for two reasons. First, the United States has ample land that allows dispersed cities to continue spreading outward. (The United States, Canada, and Australia—countries with large open spaces—own the highest number of cars per person in the world.) Second, the United States' historically low gasoline prices have not hindered people from driving places where they could walk or take a bus. Now other parts of the world have begun to follow the U.S. lead. China's expanding middle class, for example, purchases a large portion of the world's new cars, and, as a result, China now experiences the same ills caused by too many cars that the United States does. Though most people in China's 1.3 billion population do not own cars, car ownership has grown 300 percent in a mere six years. The Beijing Web site engineer Zhu Chao told the *Washington Post* reporter Maureen Fan in 2008, "I've been to Sichuan, Shandong and Jilin provinces, and I plan to spend Chinese New Year driving to Yunnan. I really like what the car brings to my life—convenience, freedom, flexibility, a quick rhythm. I can't imagine life without it." Zhu Chao could have been speaking for almost any American.

one-third the greenhouse gas emissions (gases and particles) as a truck carrying the same tonnage over an equal distance; (3) freight railroads lessen truck traffic congestion in cities; and (4) rail transport of hazardous chemicals has a better safety record than truck transport of the same chemicals. The U.S. Environmental Protection Agency (EPA) additionally stated in its 2006 report *Greenhouse Gas Emissions from the U.S. Transportation Sector* that, although transportation accounts for more than 30 percent of total greenhouse gases, locomotives contribute only 2 percent to that total.

Despite the advantages cited by the railroad industry, new rails have been difficult to build due to expense and state and local regulations. Many people oppose any new tracks running through their neighborhood. Much of the FRA's work today centers on improving the safety of existing railroads that cross places that are becoming increasingly populated. These safety precautions include the use of horns in populated areas, slowdown requirements in city limits, and improved train-traffic crossings.

The rail industry has worked to reduce locomotive emissions by attaching new *scrubbers* to the engine exhaust systems. A scrubber is a device that burns and filters soot from locomotive exhaust. In 2006 Mark Davis spoke for the Union Pacific Railroad by explaining, "It's [scrubbers] part of our industry's continued effort in looking for cleaner, more fuel efficient locomotives. We looked at 14 different filtering technologies and this one [a silicon carbide–based filter] made the most sense and best fit the rail industry's needs." Locomotive scrubbers represent an essential first step in building cleaner railroads.

Train fuel efficiency has improved more than 80 percent since 1980 according to railroad transportation company CSX Corporation. The CSX vice president Lisa Mancini praised the railroad industry in 2008, saying, "Freight rail is safe, secure, efficient and sustainable. Through dedication and cooperative investment it is possible to achieve a balance between freight rail—which drives our nation's economy—and passenger rail—which carries our nation's citizens." In order to fulfill Mancini's vision, freight transport will likely focus on the following objectives in the near future: (1) further increases in fuel efficiency; (2) equipment for reducing hazardous emissions on current locomotives; (3) development of clean-air locomotives; and (4) design of shutdown systems to reduce fuel waste and emissions when idling.

Freight transport by trucks increased dramatically as the new U.S. interstate highways were built in the 1950s. Trucks now carry 58.2 percent of goods by weight compared with 12 percent carried by train. But trucks and rail lines carry about equal amounts of goods in terms of ton-miles, meaning the miles needed to transport a ton of goods. Trucks tend to transport materials of much higher per-ton value than trains. Trains carry new cars coast to coast, but they also deliver hundreds of tons of raw materials to industry. These materials—coal, chemicals, solvents, resins—have a lower per-ton value than many of the finished products carried by trucks. Though North American transport depends on trucks, the trucking industry must confront the same issues as personal vehicles regarding emissions. Large diesel trucks contribute up to 40 percent of the nitrous oxides (a greenhouse gas) and as much as 60 percent of all particles emitted by vehicles that end up in the atmosphere. Trucking therefore needs to make a concerted effort toward cleaner alternative fuel vehicles as much, if not more, than personal vehicles.

CLEAN SHIPS

Oceangoing ships move tons of raw materials and finished products between continents, and intercontinental shipping has become more important as economies have become more globalized. Despite their role in a healthy economy, large cargo ships and oil tankers also cause air pollution and water pollution. Pollution from oceangoing ships comes from four sources: (1) exhaust; (2) ballast water, which is water that ships draw into empty tanks to provide stability and then dump out as the ship enters port; (3) fuel leaks and oil spills; and (4) dumped wastes. Laws dating back to the 1970s address air and water pollution but have ignored the wastes discharged from large ships. Residents of port cities, however, have known for some time that their air has been fouled mainly by ships. "These ships are essentially floating smokestacks," Vickie Patton, an attorney at the Environmental Defense Fund, told *USA Today* in 2004. Another concern is that ships produce more sulfur oxide and nitrogen oxide gases than CO_2, and the sulfur and nitrogen compounds contribute more to global warming than CO_2 does.

Shipping causes two additional hazards to the environment: (1) the introduction of *invasive species* and (2) water pollution. Many invasive species that have infiltrated ecosystems around the world arrived in ballast water. Hundreds, perhaps thousands, of invasive species from microbes to fish and plants have traveled via ballast water. Cargo ships, tankers, cruise ships, tugs, and smaller boats also foul waters with small but constant fuel leaks, large oil spill accidents, and accidental or intentional waste-dumping. The shipping industry therefore has two critical issues to address: (1) the need to build new clean-running ships to replace older, dirtier vessels, and (2) stricter adherence to laws against emissions, fuel spills, and illegal dumping. Unfortunately, very few laws exist to regulate ballast dumping. Governments can help by passing laws to require the shipping industry to clean up its current operations and seek new technologies to circumvent the ballast problem.

Washington's Port of Seattle has taken the initiative with a plan to require that all ships that regularly enter the port convert from diesel-powered engines to *clean ships*. This of course will be an enormous undertaking for two reasons: (1) many thousands of ships currently cruise the world's oceans, and (2) the majority of ships are registered with countries that may not make the environment a priority. Despite the challenges, the Port of Seattle has been specific in what it expects from clean ships. First, vessels in port will be allowed to plug into onshore electric power sources rather than burn fossil fuels. Second, ships may be required to burn low-sulfur fuel while at berth and perhaps while at sea. Third, the shipping industry should continue its recent push to convert to diesel blends that reduce fossil fuel combustion. Fourth, current ships can be retrofitted into more energy-efficient designs. Fifth, the efficiency of operations at terminals should be improved to speed loading and unloading and so reduce the time ships wait offshore.

As the old generation of ships disappears and a new generation takes its place, shipbuilders will pursue the following objectives, listed in order of their feasibility:

1. better shapes to reduce drag
2. low-drag paints and finishes
3. new propeller designs
4. clean-burning fuels
5. onboard emissions scrubbers
6. hull air-blasters to reduce hull-water resistance
7. hybrid vessels

In 2008 the *Christian Science Monitor* writer Gregory Lamb discussed an even more innovative approach than the previous options: "When the cargo ship *Beluga SkySails* left the port of Bremen, Germany, in January, it carried with it a high-tech version of an ancient means of propulsion. During its 11,952-nautical-mile [22,135-km] voyage to Venezuela and back, the ship launched a giant kite from its bow, sending it hundreds of feet into the air to capture the stronger and more consistent winds found above. The 1,720-square-foot [160-m²] kite, controlled by onboard electronics, exerted enough pull on the ship to provide about 20 percent of the engine power required for the journey." This approach was certainly innovative, but hardly practical for today's global shipping industry. Nevertheless, transportation always has room for people with inventive ideas to help the environment.

The Austrian engineer Michael Frauscher may have hit upon a more practical solution than huge sails. Frauscher builds hybrid recreational boats that run on electricity when starting or idling or when underway at slow speed. These boats then transition to regular fuel combustion when they reach higher speeds outside the harbor. "Everybody knows about

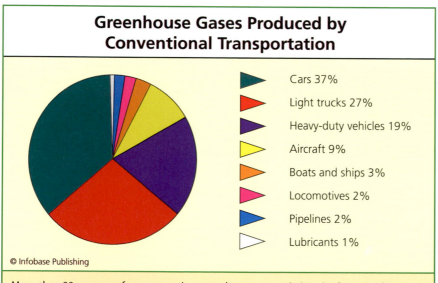

Greenhouse Gases Produced by Conventional Transportation

- Cars 37%
- Light trucks 27%
- Heavy-duty vehicles 19%
- Aircraft 9%
- Boats and ships 3%
- Locomotives 2%
- Pipelines 2%
- Lubricants 1%

© Infobase Publishing

More than 80 percent of transportation greenhouse gas emissions in the United States comes from on-road vehicles. Light trucks, or light-duty trucks, are sport utility vehicles, pickup trucks, and vans. Heavy-duty vehicles are commercial trucks, trucks with more than two axles, and buses.

hybrid cars," Frauscher said in the *San Francisco Chronicle* in 2008, "but hybrid boats are different." Since most of the pollution from boats comes when they are in harbor, the hybrid boat has been designed to first solve this problem. San Francisco has added new ferries to its existing fleet that burn a blend of biodiesel (fuel made from biological sources) and low-sulfur fuel to reduce dangerous emissions by 85 percent lower than federal emissions requirements. Hybrid technology as well as low-emissions fuels may signal the next wave in cleaner ships.

ALTERNATIVES TO TRAVEL

Any action that eliminates unnecessary travel helps build sustainable communities by conserving natural resources, but environmental engineers will likely have a hard time deciding what constitutes unnecessary travel. People travel for the following main reasons: daily commutes, business travel, and vacation travel. Business and recreational travel may be difficult to curtail because these preplanned trips affect the health of the economy. Daily commuters certainly help the economy too, but their repeated trips along familiar routes offer the best opportunity for adopting new technologies.

The transportation consultant Alan Pisarski hinted at a new way of commuting in a 2004 *USA Today* article: "Pisarski suggests that building more roads or mass transit options, such as trains or buses, would only encourage more long-distance commutes. The real change, he says, will be when companies build away from the metropolitan centers." A complete reinvention of transportation may emerge in the next 50 years. On a small scale, vehicles will still exist, so any new designs to save fuel will grow in importance. On a larger scale, the public and businesses will need to rethink how they travel and why they travel. Any major redesign in transportation for the future will emphasize the use of advanced telecommunication technologies.

Telecommuting is the process of working all or part of the normal workweek from home. Technology already exists that makes wireless telecommuting possible and increasingly common: computers, phones, hand-held communication devices, and fax machines. New technologies in *virtual commuting* can enhance the telecommuting experience for both employees and employers. In virtual commuting, a worker creates a look-alike image called an avatar viewed on a computer screen to participate in meetings with other employees. The three-dimensional images connect the worker to a main corporate office or other businesses. Virtual commuting may be able to create communities where no corporate headquarters exist; every worker resides at a remote site and communicates with others electronically. Any telecommunications process that connects people working in remote sites is called in-world meeting.

Virtual commuting enables people to sit side by side in an imaginary meeting room, give slide presentations, and communicate by speaker phone or in an online chat room. Companies retain the option to rent remote office space for occasions when employees must meet in person. Rather than commute all the way to headquarters, these employees would take shorter trips to the satellite offices and telecommute together from there to the main office.

Video conferencing offers the same conveniences as virtual commuting, but participants communicate using cameras and talk to each other via a television, projection screen, or computer screen. An A. G. Lambert executive told the *San Francisco Chronicle* in 2007, "It's about being able to have a natural experience even if you're not sitting in the same room." Sean M. Grady explained in his book *Virtual Reality,* "Virtual reality came into being in the mid to late 1980s, following decades of research into ways to remove the hardware wall between computer users and computer data." Both video conferencing and virtual commuting have a bright future when, or if, corporations redesign how they do business.

Large companies feel comfortable with traditional central offices filled with worker cubicles. The telecommunications industry must therefore make in-world meeting technologies as attractive

Telecommuting offers an excellent way to reduce greenhouse gas emissions. Engineers continue to develop innovations that reduce the sense of distance between a telecommuter and the central office. In many instances, as pictured here, a telecommuter requires uncomplicated technology: a computer, phone, fax machine, scanner, and a reliable Internet connection.

as possible to break tradition. Linden Laboratories in California designs virtual communities for businesses so that each business creates a tele-commuting system that meets its particular needs. The virtual communities take place on Second Life, a program that allows multiple users to share and edit documents, access public or private company space, and meet in virtual rooms set up to resemble the real-life conference room at headquarters. A computer industry analyst Claire Schooley noted in the *Chronicle,* "You have so much work that is done on the computer today. They [virtual technologies] make it very easy to connect." As with so many of the changes needed in transportation, breaking old habits presents the biggest challenge of all.

CONCLUSION

Transportation causes a significant amount of damage to the environment, mainly from fossil fuel consumption, pollution, and destruction of plant and animal habitat. New sustainable communities must find ways to tackle the enormous job of redesigning current modes of transportation in order to halt environmental damage. Fortunately, many opportunities await transportation engineers in the following areas: redesign of mass transit systems; personal vehicle design; redesign of truck engines to biodiesel; structures and programs to accommodate bicycling, walking, and car-sharing; and new planning and design of roads. The shipping, railroads, trucking, and airline industries also have opportunities for finding new ways to conserve energy and reduce. The airline industry has already begun to take steps to streamline their airport operations and make aircraft more fuel efficient. By comparison, the global shipping industry requires major changes to remove inefficiencies in fuel use and to stop pollution. These changes may never take place to any meaningful extent unless new and stronger international laws require that cleaner ships be built and operated properly.

Whether a customer shops for a new alternative-fuel car or a business takes the plunge into virtual reality, people must soon change their views on transportation. The old system of dirty, inefficient buses stuck in traffic next to thousands of cars—each carrying only one or two passengers—must end soon. Otherwise, pollution will accumulate, global warming will continue, and the environment will collapse bit by bit. Innovations like advanced transit offer promise in helping build sustainability by incorporating different types of transportation to suit different needs while reducing pollution and fuel waste. Advanced transit demands that transportation systems provide clean and fast vehicles that take large numbers of people where they want to go. But the expense of advanced transit can only be feasible if people realize that they must give up the freedom of traveling mainly by personal cars. This is no small task. In fact, environmental engineers may find it far easier to build new transportation systems than to convince the public to use them. Environmental engineering therefore will need help from the public, perhaps led by environmentalists and local government, to change old habits. The automobile industry also plays a key role by leading the way toward fuel-efficient vehicles rather than merely responding to people's current tastes.

Many people may believe that these challenges are too big to overcome. New transit systems require a mixture of three things in order to make advanced transit a reality. First, people in the United States need to learn that independent travel in personal cars must be drastically reduced. Walking and bicycling for short trips offer healthy options, and a greater percentage of the population should try either one, sooner rather than later. Second, the automobile industry must be willing to take the lead in making alternative vehicles the norm rather than a special subcategory. Third, businesses must accept short-term risk by saying good-bye to the old ways of managing a workforce and relying more on virtual communication. These things do not come easily or inexpensively. Environmental engineers can solve only a part of today's transportation problems. The mass transit challenge requires dedication from federal and local governments, industry leaders, and the public.

INNOVATIONS IN PERSONAL VEHICLES

Cars have become so ingrained in the culture of industrialized nations, especially the United States, that driving has become almost part of national identity. In order to use natural resources in a sustainable manner, communities will probably combine car use with other forms of transportation rather than give up cars altogether. The previous chapter discussed the role of mass transit in building energy-efficient transportation. Mass transit and the use of personal vehicles will work best if they complement each other's needs. That means well-planned, attractive mass transit running routes that take pressure off cars. But vehicles must also become more efficient by adopting design changes that greatly improve fuel efficiency and take pressure off fossil fuels. Automobile industry engineers have already made good progress in what is called designing out inefficiency, meaning redesigning any part of the vehicle that causes excess waste and extra fuel consumption.

Innovations in personal vehicles come from two directions: (1) vehicle designs that increase fuel efficiency—the distance a vehicle can travel per volume of fuel—and (2) fuels that replace gasoline and diesel and so reduce the emission of greenhouse gases. The trick in achieving success in both innovations resides in the automobile industry's ability to remain profitable while it reinvents the car. Global warming has already reached deadly levels for some ecosystems, so automakers must understand that beginning the process of redesigning personal vehicles cannot be delayed.

Why are personal vehicles so critical to the health of the environment? Cars, sport utility vehicles, and small trucks produce about 50 percent of the greenhouse gases in the atmosphere. Cars and pickup trucks in the United States contribute to half that amount even though these vehicles make up 30 percent of all the world's cars and pickup trucks. These vehicles produce the greenhouse gases nitrogen oxides, sulfur dioxide, and carbon dioxide, which all trap heat in the Earth's atmosphere. This global warming has caused, and continues to cause, plant and animal species to disappear because they cannot adjust to habitats that have changed due to warmer temperatures. Warming alters the plant life and prey-predator relationships in ecosystems; food sources disappear; invasive species enter ecosystems and destroy them. Vehicles, industries, and residential buildings all contribute to this warming effect.

This chapter explores aspects of vehicle design that can help reduce environmental harm. It emphasizes alterations to car design that would make each vehicle less thirsty for fossil fuel and produce less hazardous emissions. It also describes special topics such as aerodynamics and drag and surface technology, which engineers study to find the best innovations for making car design more energy efficient. Two special sections discuss today's automobile industry and the U.S. interstate system's benefits and drawbacks.

NEW VEHICLES EMERGE

One reason why energy-efficient cars have not dominated the automobile market originates in resistance to change, either by manufacturers or by consumers. Small minor changes always gain acceptance faster than drastic changes, and so the world's push toward environmentally sensible cars has come in baby steps. An increasing number of drivers in North America and Europe now accept new electric cars and gasoline-electric hybrid cars because they know these models cause less harm to the atmosphere than conventional combustion-engine cars. But these alternative choices remain only a small proportion of the total vehicles on roads today—the automobile industry continues to refer to alternative vehicles as a "segment" of the industry.

Automotive engineers have a formidable challenge ahead of them: to transform alternative vehicles from a novelty choice to the main type of vehicle car buyers want. In time, alternative vehicles may become the only type of vehicle on the road. To do this, engineers must combine their skills

Alternative vehicles that rely on fuels other than gasoline or diesel fuel have become increasingly numerous on public roads, but they still make up a small percentage of total vehicles. Ethanol-powered cars comprise almost half of all alternative vehicles; about one-quarter run on liquefied gas. This Honda at a hydrogen filling station may represent the next generation of alternative fuel vehicles. *(National Hydrogen Association)*

in developing new ways to power vehicles with designers' skills in making the vehicles appealing to buyers.

A new generation of vehicles will contain one or more of the following features: an alternative fuel engine; aerodynamic design; a frame composed of all recycled materials; and smaller size. The best opportunity to see examples of new vehicle designs occurs at international auto shows. Ron Cogan, the editor of the *Green Car Journal,* said in 2007, "The international car shows are a window to what the designers are thinking, and I began seeing actual designs for efficiency." Cogan and other car experts have been hoping for a *zero emission vehicle* as the first major breakthrough in car redesign since the hybrid vehicles emerged in the 1990s. Cogan said of the zero emission vehicle, "No one knew it could be done until General

Motors unveiled 'Impact I.'" The zero-emission Impact I appeared in early 1996, but shifting state emission laws complicated the Impact I's rollout and it never became a marketable car for the public. In other words, the car's engineering succeeded and its design may well have succeeded, but complicated pollution laws scuttled its future.

Automotive engineers and car designers can refer to Toyota's Prius to learn how a radically new car concept can become a success. The Prius was the first highly marketable *green car,* a car designed to cause little or no harm to the environment. The Prius arrived in 1997 with an engine containing components that used either gasoline or electric power, and the vehicle switched back and forth between these two energy sources in a single trip; this new idea was called a *hybrid vehicle.* A hybrid is any vehicle that uses more than one type of energy source in converting fuel into motion. Several car manufacturers have since built their own hybrid models, and consumers have been enthusiastic toward giving up conventional vehicles

Car designers and engineers have tried to develop alternative fuel cars that resemble conventional combustion engine cars. By designing according to a traditional appearance, new models will appeal to car buyers who want fuel savings but are reluctant to buy unusual-looking cars. This 2008 Honda Civic GX uses compressed natural gas as fuel and can refuel from a specialized appliance kept at home. *(Honda Media Newsroom)*

for them. Jim Motavelli wrote in 2008 for *E, the Environmental Magazine*, "[Toyota] has 80 percent of the U.S. hybrid market, and the majority of those sales are the brilliantly designed Prius. The four-door Prius caught on both because of its incredible fuel economy—48 miles per gallon [77 km/h] in the city and 45 miles per gallon [72 km/h] on the highway—and because its unique styling makes an environmental statement about its owner. It *looks* like a hybrid." The latest green cars have built upon the principles of the Prius and incorporate many of its technologies.

Automakers produce concept cars to test innovations in fuel use, aerodynamics, materials, and design. These models are not ready for highway use, but the innovations they possess might become part of cars in the future. The car pictured here is the Aptera made by Aptera Motors, Inc. This model is completely electric. The company has also developed an electric/diesel hybrid Aptera, which gets 330 miles per gallon (140 km/l). *(Aptera Motors, Inc.)*

Most hybrid vehicles like the Prius still depend to some extent on gasoline. Fuel efficiency in these cars relates to the means in which they conserve gasoline. Next-generation green cars, however, will move beyond gasoline altogether. Ron Cogan pointed out, "If you look at the big picture, at some point we will exit from oil." New cars will combine alternative-fuel engines with designs that maximize energy-use efficiency. The table on page 72 describes some of the prototypes (also called *concept cars*) that car manufacturers have unveiled in an effort to create a truly new car. Whether these ideas succeed or fail may not be as important as the fact that big automakers were willing to attempt completely new car designs.

Getting prototypes off the drawing board and onto the road requires co-operation among manufacturers, suppliers, retailers, and government. The Green Car Congress has stated, "The path to sustainable mobility is complex, with numerous competing and complementary approaches to alternative energy sources, production, distribution and applications; fuel and power-train

EXAMPLES OF GREEN VEHICLE PROTOTYPES

MODEL	DATE	MAIN FEATURES INTRODUCED
DaimlerChrysler NECAR	1999	hydrogen fuel cell–powered
REVA	2001	battery-powered electric
Honda Puyo	2007	seamless contoured body lacking sharp corners; variable color in which car turns green when it is running at peak efficiency
Renault ZE	2008	electric car with surface that decreases temperature fluctuations
Fiat Siena Tetrafuel	2008	first bi-fuel car, runs on gas or compressed natural gas
Carver One	2008	drives like a car but tilts like a motorcycle on curves
General Motors Volt	2008	battery powered for extended distance
Nissan Nuvu	2008	all-electric power with solar panels on the roof
Peugeot RC HYmotion4	2008	aerodynamics maximized; carbon body on aluminum frame
Honda Civic GX	2008	compressed natural gas–powered that can be refueled at home

options; materials; safety, economic and environmental considerations and trade-offs; policy issues; and different time lines for research, development and deployment." The role of carmakers is explored further in the following sidebar "The Automobile Industry."

DRAG AND ENERGY LOSS

The Ford Model T had little need for aerodynamic styling when it rolled onto the road in 1908; it never ran faster than 45 miles per hour (72 km/h). Today's high-performance cars reach 180 miles per hour (290 km/h), although drivers on open highways usually travel closer to 70 to 90 miles per hour (113–145 km/h). A Model T's blocky frame would not withstand today's speeds even if it were engineered to travel that fast. Cars today have designs that reduce wind resistance and drag, both of which cause a vehicle to use more power and more fuel to maintain speed.

Aerodynamic drag consists of forces that work against a vehicle's forward motion. Examples of things that increase a car's drag are the following: shape, area, density, speed, incline, viscosity (density of the air), and surface friction. Aerodynamic drag accounts for about 20 percent of the energy consumed by most car models. For this reason engineers study drag in all new vehicle designs to optimize energy efficiency.

All cars, ships, boats, and airplanes have been designed based on a *drag coefficient* (Cd), which is a unitless value used to describe the aerodynamics of a vehicle in motion. Put another way, Cd describes the relationship between the air's motion and a vehicle's motion. Car drag coefficients range from 0.25 to 0.45, with sleek Corvette-style cars on the low end of the Cd range and boxy Hummers at the high end. Usually only the newest prototypes achieve a Cd of 0.25, although the Prius measures 0.26. Car designers and engineers reduce drag and therefore enhance fuel efficiency by the following actions:

1. remove sharp corners from the body design
2. reduce the frontal area
3. remove extra components such as spoilers and roof racks
4. avoid wide tires
5. increase windshield angle
6. remove high axle to lower the car's body

The best way to reduce drag in a well-designed vehicle is to reduce driving speed. Cars use up most of their energy overcoming the effects of drag that increases disproportionately at higher speeds, meaning a small increase in speed causes a large increase in drag. Car designers have used race cars as examples of low-drag design. Race car design incorporates two critical factors for achieving high speeds with a minimum amount of drag: stability and downforce. At speeds of more than 200 miles per hour (322 km/h), race cars must have adequate stability and downforce, which keeps the car from going airborne, a situation called *lift*. Passenger vehicles travel at speeds that remove the risk of lift, and they remain stable at legal speed limits. Auto industry engineers nevertheless consider stability and downforce with each new design.

THE AUTOMOBILE INDUSTRY

The automobile industry consists of a massive enterprise that plans and designs cars and light trucks and produces, distributes, markets, and sells these products. The industry sells about 75 million cars worldwide in a year; the United States purchases more than 11 million cars each year. The United States also produces about 25 percent of all cars made worldwide. Only Japan makes more cars than the United States, but China has been closing the gap fast and may soon overtake U.S. carmakers. Car manufacturing supports many other industries, primarily the oil industry. Makers of engines, plastics, metals, fabrics, leathers, tires, and small components of car bodies, plus recyclers, service stations, and businesses that install sound systems and other add-ons all depend on the car industry for their income. The automobile industry, which controls the development-to-sale process, combines with all the other associated industries to form the more all-encompassing automotive industry.

The first automobile built to carry passengers arrived in 1801 due to the inspiration of the British inventor Richard Trevithick. Through the 1800s, vehicles self-propelled by steam engines became common. The French inventor Étienne Lenoir and the German engineers Eugen Langen and Nicholaus August Otto worked on building a combustion engine to run on fuel and replace steam, a move intended to make cars more powerful. By 1876 Otto developed a four-cycle combustion engine that would be the precursor to today's engines. In 1913 the American Henry Ford invented a conveyor belt to take components directly to workers on a car assembly line, and from this point onward, American automakers would dominate the automobile industry for the next 75 years.

The development of the combustion engine and Ford's innovative mass production promised a bright future for the automobile industry. The U.S. automobile industry dominated world markets until the 1980s when Japanese automakers mounted a strong challenge to American

EFFICIENT VEHICLE DESIGN

The first stage in new vehicle design begins with research on the needs and desires of drivers, government regulations on emissions, and new technologies for improving performance. Each of these facets must also align with the intangible aspect of how a car's looks will appeal to car buyers. Obviously, new car design involves large teams of experts in all of these specialties who must coordinate their work in a way that results in a feasible prototype car.

models. In 1970 the Clean Air Act brought new laws for car emissions, but the automobile industry learned to sidestep most of the regulations with a few minor concessions, such as the catalytic converter that made combustion more efficient. (Catalytic converters also caused a switch to lead-free gas because the devices did not work on leaded gas.) The act's 1990 amendments forced automakers into taking further steps toward inventing clean-running engines that produced less hazardous emissions than earlier models. Today the automobile industry continues to adjust to tighter restrictions on exhaust emissions, led by California with the strictest emissions laws.

An Arab oil embargo in 1973, fuel crises throughout the 1970s, and rising fuel prices in the 2000s have been events that put pressure on the automobile industry to find more avenues toward fuel efficiency. In all these instances, however, automakers reacted to new requirements rather than led the way in innovation. In 2009, a faltering world economy put pressure on automakers to accept President Barack Obama's new fuel legislation. "They [automakers] can feel the political winds changing," said lawyer David Doniger of the Natural Resources Defense Council. "They need government aid to stay in business. When you have your hand out for help, it's hard to use the same hand to thumb your nose at the federal government."

Temporary fuel crises certainly affect the auto industry and car sales, but a more overwhelming threat to the industry's conventional way of doing things comes from deep underground. The crude oil that the oil industry taps and turns into fuel will not last forever. The auto industry cannot wait until the last drop of oil comes out of the Earth before it reacts. Mike Millikin, the founder of the Green Car Congress, and the environmental writer Alex Steffen wrote in 2006, "The auto industry was built on a seemingly endless supply of gasoline, but it is now becoming increasingly clear that the end is, in fact, in sight." The end of oil should not signal the end of the auto industry, but to avoid going extinct this industry's leaders must plan for the future now.

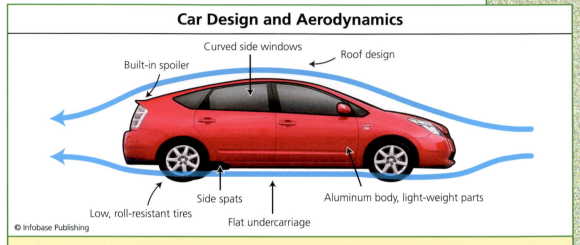

Car Design and Aerodynamics

Built-in spoiler

Curved side windows

Roof design

Low, roll-resistant tires

Side spats

Flat undercarriage

Aluminum body, light-weight parts

© Infobase Publishing

Car designs include features to reduce drag as they move. Many of the features such as window shape, roof shape, and rear spoiler help airflow over the car. Side spats and a mostly flat undercarriage reduce pockets where air can swirl and decrease the car's aerodynamics. Vehicles also contain lightweight materials on their surfaces that allow smooth airflow over the car's body.

Rather than depend solely on drawings—even sophisticated computer drawings—automakers build prototype models of all of their most promising car designs. Building a prototype car is a complex job that requires a skilled team. Prototype builders follow a design blueprint to create a full-sized fiberglass shell, windshield, and lights. Computer-aided design (CAD) has greatly increased prototype building by allowing for design changes on the computer before building the actual car mock-up. CAD programs also create images of the vehicle's interior controls, seats, and options. Engineers use CAD programs to make up-to-the-minute design changes, and then the prototype team adjusts the model to agree with those changes. The entire process is called virtual prototyping, and, in addition to car designers, trucks, aircraft, ships, pleasure boats, and mass transit vehicles use it. This technology has helped increase the accuracy of converting a concept into a real design that will succeed on the road.

AERODYNAMICS

Car designers must satisfy customers' tastes while at the same time incorporating new technologies that improve aerodynamics. Aerodynamics is the physics that enables things to move through the air with minimal resistance, also called aerodynamic drag or wind resistance. Designers shape the outer body of cars to get the best aerodynamics and also design the interior to minimize the wind noise a vehicle creates when moving at high speed. The aerodynamic models sold today take into account both exterior and interior reactions of the vehicle with the air it moves through. The automobile industry uses computer modeling and wind tunnel testing to achieve the new model's desired performance.

Aerodynamics engineering for cars focuses on two main components of drag: ground effects and rolling resistance. Ground effects are design features that reduce lift and wind resistance. Car designers shape a car so that more air flows smoothly over the car's top than flows across the underside. Such a design also includes structures that reduce the turbulence of the air moving over the car and off the tail. In most vehicles the junction between the roof and rear window contributes to turbulence. For this reason new fixtures have been introduced to cut this turbulence and improve airflow, a process engineers call "tripping the airflow." The following components trip the airflow on various new car models:

1. turbulator strips—thin metal strips that lessen the small layer of turbulence that occurs between any surface over which air flows

2. vortex generators—small teeth attached to the roof above the rear window to even the airflow speed over a vehicle

3. diffusers—undercarriage components that direct airflow away from the wheels to increase a downward force on a vehicle

4. shortened fairings—body parts that direct airflow around the tires

POWER

Power is the rate in which energy converts into work. Combustion vehicles use mechanical power produced by the engine, and this power comes from the engine's force and velocity:

$$power = force \times velocity$$

Alternative fuel cars, such as gasoline-electric hybrids, must possess features that maintain power while reducing fuel use. These attributes enable alternative fuel cars to compete with traditional gasoline cars.

The engine supplies power, and the vehicle moves. The engine's cylinders burn fuel, usually gasoline, in a process called combustion and convert the fuel's energy into motion (the turning of the engine's crankshaft). Regardless of the type of fuel a vehicle uses—gasoline, diesel, electricity, a fuel cell—carmakers express the vehicle's power as horsepower, which is a unit that describes the rate at which work is done. One horsepower equals 746 watts and stands as an internationally recognized unit to measure power. (Engineer James Watt came up with horsepower in the 1700s based on his interpretation of how much work a horse could do in one minute. By Watt's calculations, one horse could pull 330 pounds [150 kg] of matter 100 feet [30 m] in one minute: 1 horsepower = 33,000 foot-pounds per minute.) A typical vehicle uses 20 horsepower to travel at 50 miles per hour (80 km/h) for one hour. Today's combustion engines produce 120 to 270 horsepower, a range that gasoline-electric hybrid cars and the newest prototypes also attain.

It may be hard to decide whether powerful cars preceded superhighways or superhighway systems led to more powerful cars. The following "Case Study: U.S. Interstate Highways' Effects on the Environment" examines an often overlooked component of transportation.

VEHICLE SURFACE TECHNOLOGY

In addition to aerodynamic shape, lightweight components, and optimized engine efficiency, car designers can improve a car's overall efficiency by modifying the vehicle's surface. Moving objects have drag due to skin friction, which is the interaction of a surface with the air moving over it. Auto industry engineers therefore must reduce *skin friction drag* in order to increase fuel efficiency. The rougher a vehicle's surface, the more drag the vehicle experiences. By eliminating even microscopic rough spots in a car's paint coating, engineers help conserve fuel.

New coatings that cover a vehicle's paint have been investigated to reduce skin friction drag. For example, thin plastic coatings overlaid on paint can affect the roughness of a vehicle's surface and so affect drag. The best coatings create a smooth surface for airflow and resist scratches and corrosion, causes of surface roughness. A smooth surface then allows air

Case Study: U.S. Interstate Highways' Effects on the Environment

The U.S. system of cross-country highways came into being when President Dwight Eisenhower signed the Federal-Aid Highway Act of 1956. On June 30 of that year the *New York Times* reporter John Morris wrote of the event, "President Eisenhower set into motion a record $33,480,000,000 road-building today by signing the bipartisan authorization bill that Congress sent him Tuesday. Sinclair Weeks, secretary of commerce, immediately announced the allocation of $1,125,000,000 among the states for the first year of what he called 'the greatest public-works program in the history of the world.'" Secretary Weeks may have overstated the magnitude of the 42,000-mile (67,592-km) highway network to be built, but the interstate system nevertheless made an impact on lifestyles and businesses for years to come.

The U.S. interstate highways provided a different type of roadway from state roads and local thoroughfares because the entire network contained only about 16,000 interchanges that drivers use for entrance or exit. This design made travel more efficient for drivers by avoiding traffic lights and busy intersections. As the highways reached across the nation throughout the 1950s and 1960s, they also began to exert a subtle effect on the environment.

The interstates—north-south routes have increasing odd numbers from west to east; even-numbered east-west routes increase from south to north—made personal travel, business trips, and trucking more convenient and faster. The interstates improved travel so much in fact that they contributed to urban sprawl from city centers; densely populated suburbs have grown up near many interchanges. Though the interstate highways made life easier for communities and commerce, they also led to the following environmental issues:

1. car and truck exhaust
2. animal and plant habitat destruction
3. animal and plant habitat fragmentation

(continues)

(continued)

4. blocked wildlife migration corridors

5. wildlife road kills

6. noise

7. high fuel consumption

In addition to the environmental ills mentioned here, the interstate highways have probably helped build the U.S. car culture in which more people depend on personal cars than on mass transit. The highways surely changed the way American society matured between the 1950s and 1970s as people became more mobile and the country seemed to become smaller. (Interstate highways also gave birth to the fast food restaurant industry.) The *National Geographic* writer Robert Paul Jordan said of the interstate system at the height of its construction, "Americans are living in the midst of a miracle. A giant nationwide engineering project—the Interstate Highway System—is altering and circumventing geography on an unprecedented scale." Jordan's description captured the importance of today's interstate highways, but the highway system's progress also brought with it a decline in the health of the environment.

to pass over the vehicle in laminar sheets, meaning that streams of air line up in parallel sheets as they move. Turbulent air, by contrast, contains eddies and swirls that cause skin friction drag.

Engineers study the boundary layer of air just next to a vehicle's surface to determine why most of the boundary layer contains laminar flow but some parts contain turbulence. The German physicist Ludwig Prandtl introduced the idea of boundary layers in 1904 and explained that the friction between a moving object and air (or water) developed only in the thin boundary layer that existed very close to the surface. Scientists since Archimedes (287–212 B.C.E.) have pondered the way flow moves over and around solid objects, but Prandtl's boundary layer theory proposed that all flow does not behave the same in the presence of a surface. Prandtl explained that flow nearer a surface moves slower than flow farther from the surface: "A very satisfactory explanation of the physical process in the boundary layer between a fluid [or airflow] and a solid body could be obtained by the hypothesis of an adhesion of the fluid to the walls . . . If the viscosity [thickness of a fluid] was very small and the fluid path along the wall not too long, the fluid velocity ought to resume its normal value at a very short distance from the wall. In the thin transition layer however, the sharp changes of velocity, even with small coefficient of friction, produce marked results." In other words, air sheets close to the solid surface move at slower speeds and contribute to skin friction drag more than the airflow outside that boundary layer, which moves at normal speed.

Surfacing engineering now focuses on substances to increase airflow at the boundary layer. Engineering experiments have discovered that a little roughness at certain points along the boundary layer actually helps with airflow by weakening the boundary layer–surface association. For this reason, some cars contain rough outer body components just under the doors and below the radiator grill. Specialists in surface science now study the structure of sharkskin, which reduces a shark's drag and allows it to speed through the water. Sharkskin contains a series of tiny rigid ridges called *riblets* that feel rough to the touch but greatly reduce drag in

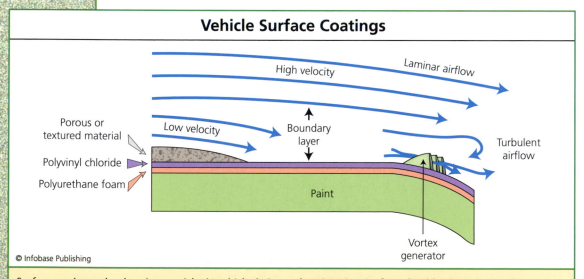

Vehicle Surface Coatings

© Infobase Publishing

Surface coating technology is a specialty in vehicle design and engineering. Surfaces should reduce drag and assist airflow. All sources of turbulent air movement, which decrease aerodynamics, must be eliminated or corrected. The vortex generator is a simple row of fins that smooth the airflow off the back of a vehicle where turbulence is common.

water by interacting with the skin-water interface. The German engineer Volkmar Stenzel has been creating lacquers for vehicles and watercraft based on this *riblet effect,* sometimes by building the riblet structure into the lacquer itself. Stenzel said, "Our trial lacquer is based on the chemistry used in aviation paints." Surface engineering to improve the aerodynamics of a moving vehicle therefore remains an intricate science that borrows from technologies used in aircraft and watercraft. New vehicles will require the best combination of smooth surfaces, flow-direction surfaces, and vehicle shape.

CONCLUSION

New vehicle design can be considered one of environmental engineering's most promising opportunities to reduce the use of fossil fuels and lessen the amount of hazardous emissions that enter the atmosphere. New car designs therefore focus mainly on alternative fuels and on designs that improve the car aerodynamics so that the car conserves fossil or nonfossil fuels. Innovations in personal vehicles have concentrated on engine type, vehicle shape, and surface materials to achieve these objectives in fuel efficiency. Several prototype vehicles emerge from the automobile industry each year with improved aerodynamics and technologies that reduce the drag put on the car by wind resistance and rolling resistance.

Engineers and designers possess the skills to reinvent today's personal vehicles for better fuel use, but other factors make it difficult to market the new vehicles to the public. First, car buyers must accept new designs that bear little resemblance to cars presently on the roads, and, second, the new cars must perform with the same power and speed that drivers want. Third, the large automotive industry must be willing to take risks in leading with new products rather than following consumers' current tastes in cars. Stricter laws on car emissions can certainly push automakers and the public toward cleaner cars, and, despite the hesitancy that many people will probably have toward change, one car, the Prius, has already shown that a new generation of drivers is willing to make a change to benefit the environment.

The U.S. interstate highway system and other national highway systems have strengthened the car culture. Because of this, environmental engineers must realize that personal vehicles will always be a big part of industrialized society. Rather than get everyone out of their cars and onto mass transit, a better plan would be to encourage more efficient and enjoy-able mass transit experiences and then supplement that mode of travel with a new generation of cars that demand little from the environment and put little harmful waste back into the environment. Aerodynamics, new energy sources, and surface technology offer good opportunities for continued innovations that may someday make this plan a reality.

SUSTAINABLE MANUFACTURING

Sustainable activities have become commonplace in the workplace in regard to switching off lights, conserving air-conditioning and heating, and recycling. A much larger opportunity in natural resource conservation and sustainable operations, however, resides in the manufacturing sector.

Large manufacturing plants have several opportunities where they can contribute to better treatment of the environment because of the following characteristics: (1) large amounts of natural resource and raw material consumption; (2) heavy energy demand; (3) dependence on hefty equipment and heavy modes of transportation (trucks, railroads, shipping); and (4) large amounts of hazardous and nonhazardous wastes.

The world's economies use three types of resources to produce goods and services: (1) natural resources; (2) manufactured resources; and (3) human resources. Natural resources are clean water, clean air, land, forests, crude oil, coal, minerals, and radioactive elements. An easy way to define natural resources is any matter produced by Earth without human input. Manufactured resources are things that humans make, usually products or materials that come out of manufacturing facilities such as factories, mills, and refineries. Typical examples of manufactured resources are consumer products, lumber, textiles, and petroleum products such as gasoline. Human resources are the people who work in support of the economy.

Sustainable manufacturing is any activity that minimizes the use of natural resources in the process of making a product. Australia's

Commonwealth Scientific and Industrial Research Organization (CSIRO) has explained that sustainable manufacturing encompasses "technologies and engineering practices to enhance the productivity of resources and minimize environmental burdens." Therefore, in order to conserve natural resources, industries must find ways to be more efficient in making manufactured resources and managing human resources.

Sustainable manufacturing, also referred to as green manufacturing, does not come without extensive planning, some increased expense, and the natural tendency of the business world to resist change. Challenges to sustainable manufacturing make this one of the more complex aspects of environmental science. Some companies have taken steps toward sustainable resource and energy use, however, so a blueprint does exist for other companies to follow. In time, the most innovative companies may achieve zero energy and/or *zero discharge* operations in which a manufacturing plant produces all of its own energy and recycles materials so that it puts no waste into the environment.

In order to achieve zero energy and zero discharge conditions, manufacturing companies have numerous options. This chapter discusses the main methods for converting current manufacturing practices into less environmentally destructive practices. The chapter covers the structure of today's manufacturing plants, the wastes that come from large plants, pollution control, and energy management in manufacturing. It also discusses an area of concern for many companies—how to balance decisions that are good for the environment with the need to make a profit. It also shows the ways in which manufacturers can lead programs toward sustainability and benefit entire communities in the process.

TODAY'S MANUFACTURING PLANTS

A manufacturing plant, sometimes called simply a factory, has become synonymous in many people's minds with environmental decay. Factories seem to be the source of disturbing amounts of choking emissions, toxic chemicals, and polluted water. Of course, factories also produce the goods that make life easier for millions of people around the globe: life-saving drugs, foods, personal care products, household products, and electronics.

Factories originated several hundred years B.C.E. in places that experienced fast population growth such as China. Factories became commonplace in communities in need of large quantities of specific products.

Manufacturing has made strides in improving efficiency, but it needs to improve its sustainable practices, such as waste reduction, recycling, and factories constructed of ecologically friendly materials. This motorcycle production line may be using procedures that have changed little in the past century. *(Royal Enfield Motorcycles)*

Early manufacturing plants ran on human muscle power with a small amount of mechanical help. Manual labor dominated factories through the 1800s until the auto designer Henry Ford revolutionized manufacture by designing mass production. The steam engine, power tools and machinery, and mass production coalesced to reinvent manufacturing soon after Ford's first assembly line came into being. By the beginning of the 1900s, the Industrial Revolution hit its stride using these engineering innovations. The speed and the efficiency of manufacturing would grow in a way never before seen in history.

A typical manufacturing plant today contains the following components that work together to make production as efficient, inexpensive, and fast as possible:

1. production floor(s), also called the assembly line
2. raw material storage facilities
3. finished product warehouses
4. shipping and receiving sites
5. management and accounting offices

A collection of resources in addition to factory workers keep the plant running day after day; in a healthy economy, manufacturers may run three shifts that allow round-the-clock production seven days a week. Manufacturing plants consume massive amounts of the following resources: water, air, land, energy in the form of oil or coal, and raw materials such as metals, wood, plastics, or fabrics. Some plants do not rely on their own oil- or coal-produced energy, but draw electricity directly from the community's power supply.

Designers of newer factories have attempted to streamline the traditional format described here for the purpose of saving costs on energy and resources. In order to reduce the costs of storing tons of raw materials and warehouses full of finished products, modern manufacturing has adopted *just-in-time* production, which minimizes the amount of materials that must be stored at any one time by the company. In just-in-time manufacturing, raw materials come to the factory only in the amounts needed for the production of a new order. The new batch of finished products then enters a *distribution chain* immediately, usually by truck, and goes directly to the retailers that placed the order. Just-in-time manufacturing helps conserve resources in the following ways:

1. less land needed for large warehouses and storage
2. efficient distribution routes that conserve fuel
3. efficient operations on the production floor to save plant energy demand
4. less wasted raw materials
5. less excess, unsold finished product

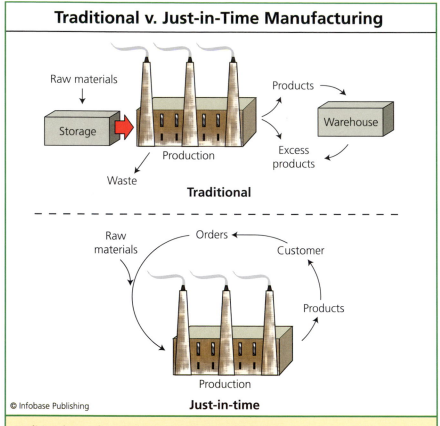

Traditional v. Just-in-Time Manufacturing

Traditional

Just-in-time

© Infobase Publishing

Traditional manufacturing relies on large production runs, warehousing of finished products, and management of excess materials and waste. Just-in-time manufacturing consumes raw materials and makes products only as needed and reduces storage to save energy. Many computer makers now use just-in-time manufacturing. Technicians assemble the computer and its components only at the time of each online order.

These innovations streamline manufacturing to conserve resources but also make businesses more profitable. In fact, innovations such as just-in-time production were planned mainly for better business and not necessarily to help the environment. Fortunately, many decisions that make a business more profitable can also spare natural resources and energy. Today's new plans for sustainable factories put a good deal of emphasis on waste reduction and energy conservation, which both help contribute to profits.

WASTES AND EMISSIONS

Manufacturing produces both solid and liquid wastes that must be treated or disposed of so they do not harm the environment. Traditional manufacturing plants produce large quantities of wastes from the following: production scraps, defective products, overproduction, or excess raw materials. Manufacturing additionally produces process wastes, which are materials needed to make a product but that do not become part of the final product. Concentrated acids are common process wastes that must be managed so they do not enter the environment where they would cause immediate damage to ecosystems. Other common process wastes are water, solvents, salts, and metals. The U.S. Environmental Protection Agency (EPA), which oversees the nation's waste management, has promoted *lean manufacturing* as a way to reduce these solid and liquid wastes. The table below summarizes the features of lean manufacturing.

An important initial step for manufacturers to take in reducing solid, liquid, and gas wastes is to list all of the materials exiting the factory as

CHARACTERISTICS OF WASTE MANAGEMENT IN LEAN MANUFACTURING	
WASTE TYPE	**LEAN MANUFACTURING PROCESS TO REDUCE THE WASTE**
scraps	return to raw material stock, reuse in other processes, transfer to another industry for recycling
defects	improved product design and production equipment design to reduce defective products
overproduction	adjust production output to product orders
raw materials	return to raw material stock, sell to another industry
process	alternative production methods, transfer to other industries

waste. Waste management begins by knowing exactly what waste materials are being produced, the amount of each, and the potential hazard of each waste on the environment. With this list in hand, production engineers follow a stepwise process to reduce the amount of these wastes produced, as follows:

1. Classify all wastes according to hazards they cause in the environment.
2. Identify the wastes produced in the highest volume.
3. Design new production processes to reduce these first, from the most costly waste to the least costly.
4. Identify the wastes produced in lower volumes or intermittently.
5. Study the production process to find ways of reducing the amount of low-volume or intermittent wastes.

Production engineers often create loops in the production process so that high-volume wastes may substitute for low-volume wastes, thus eliminating one waste category and reducing the other. The DuPont Company described a similar process it has used for several years to reduce the wastes that have cost the company the most money. By asking and finding answers to the following questions, companies like DuPont can blend environmental needs (waste reduction) with business needs (lowered costs):

- How can we reuse high-volume wastes?
- How can we modify the production process or the chemistry to reduce or eliminate each waste on the list?

The Sharp Corporation headquartered in Japan has devised a similar point-by-point plan for its electronics manufacturing plant in Kameyama that addresses several waste issues as well as energy savings. Sharp's plan comprises the following points:

1. Minimize greenhouse gas emissions.
2. Reduce resource consumption.

3. Reduce energy consumption.

4. Manage and reduce waste discharges.

5. Evaluate health risks of accidental spills of wastes, to humans and to environment.

6. Evaluate the *environmental burden* of specific solid, liquid, and gas wastes.

Some of the Kameyama innovations consist of reusing excess heat energy made by the plant's production processes to supply power for the air-conditioning system. The principles of heat energy are further discussed in the sidebar on page 94 "Heat Energy." In this case heat represents yet another manufacturing waste that can be turned into a reusable resource. Even the innovative Sharp engineers have a difficult time reusing emissions that contain carbon dioxide (CO_2) among other gaseous wastes, so for the present, manufacturers depend on emissions-cleaning technology such as scrubbers. Future production processes may someday be entirely redesigned to use biological rather than chemical processes and so eliminate dangerous emissions. This new biology-based manufacturing is called *white manufacturing,* or white biotechnology.

Traditional manufacturing often uses coal-fired plants to supply the power to run operations. These types of power plants have been linked to the worst emissions that manufacturing has produced since the dawn of the Industrial Revolution. Coal-fired power plants produce electricity by burning coal in a boiler to heat water into steam. The pressurized steam then enters a turbine that runs a generator, which in turn converts kinetic energy (motion) into electricity. These power plants use up large amounts of coal—a plant near Knoxville, Tennessee, burns 14,000 tons (12,700 metric tons) daily—and release a large volume and variety of greenhouse gases, heavy metals, and particles unless their smokestacks have scrubbers. The energy expert Gordon Couch explained to *National Geographic* in 2006, "When you try to burn coal or convert it to something else, you've got to deal with pretty difficult mineral matter. You've got sulfur, pyrite, quartz, silica and all kinds of stuff in with the coal." Power plants must maintain and clean their scrubbers on a regular schedule in order for these devices to clean hazards out of flue gases.

Coal-fired plants have two options for solving their emissions problems: redesign the manufacturing operations to use alternative energy

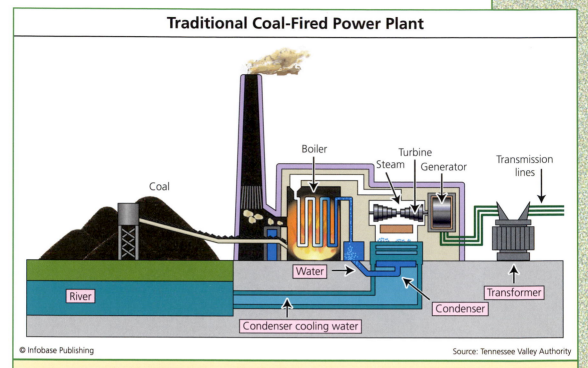

Traditional Coal-Fired Power Plant

Coal

Boiler

Steam

Turbine

Generator

Transmission lines

Water

River

Condenser cooling water

Condenser

Transformer

© Infobase Publishing

Source: Tennessee Valley Authority

Coal-fired power plants in the United States and many other parts of the world generate electricity for millions of people. These plants consume fossil fuels, water, and energy, and they produce emissions, waste ash, and hot water that harm ecosystems when released. Power generation has two options to improve the situation: design cleaner coal combustion or turn to renewable sources for electricity generation.

sources, or convert to the burning of *clean coal,* also known as low-sulfur coal. Clean coal is equivalent to standard coal with an exception: the burning of clean coal includes various steps and technologies that remove much of the particles, heavy metals, and the greenhouse gases sulfur dioxide and nitrogen oxides. The following table lists technologies that convert traditional coal-burning power plants into clean coal plants.

Critics of clean coal remind the public that the hazards of coal do not disappear by inventing technologies to lower emissions. Coal mining remains a hazardous occupation for miners, often causing environmental diseases such as black lung disease. In 2007 Marilyn Snell wrote on behalf of the Sierra Club that coal is "dirty and destructive: entire mountaintops are removed to get at it; emissions from coal-fired plants contribute to at

CLEAN COAL TECHNOLOGIES	
TECHNOLOGY	**EFFECT**
wet scrubbers	sprays flue gases with limestone mixed with water, which removes acid-forming sulfur dioxides
dry scrubbers	filters remove particles and metals from flue gases
burners	burns excess oxygen during combustion, so reduces the amount of nitrogen oxides produced
electrostatic precipitators	removes particles from flue gases by imparting an electrical charge on the particles and then collecting them on a charged plate
gasification	coal decomposes under pressurized steam and air to form gases carbon monoxide and hydrogen, which feed a gas turbine to produce electricity
supercritical burning	extremely high temperature (greater than 705°F [374°C]) and pressure (greater than 3,212 pounds per square inch [psi]) converts coal to a liquid-gas substance and increases efficiency with less waste

least 24,000 premature deaths a year in this country [the United States] alone; and it accounts for 36 percent of our overall releases of carbon dioxide, the main culprit in global warming. Despite the industry's hype, there's no such thing as 'clean coal.' But new technologies and policies can help reduce coal plants' deadly emissions, including carbon dioxide, sulfur dioxide, mercury, and nitrogen oxides." The question surrounding clean coal may be centered more on the willingness of the world to accept a costly new technology to replace the relatively cheap price of dirty coal.

Will the coal industry as the world knows it ever disappear or at least transform itself into a cleaner form of energy generation? Places in the United States have made efforts to wean themselves from coal-generated energy. California in 2007, for example, announced rules that would bar any new municipal utility company from signing a contract with a coal-fired power plant. But coal costs one-sixth the price of natural gas and a fraction of today's fluctuating oil prices, and the United States holds the world's largest coal reserves. The rest of the world uses coal as its main energy source, especially the fast-growing economies in China and India.

The coal industry will likely survive any attempts to dismantle it. But this industry can help the environment by voluntary means or through strict air pollution laws. The coal industry can pursue clean technologies, low-impact mining methods, or an advanced technology called coal-to-liquid in which coal turns into liquid *hydrocarbon* fuel under extreme heat and pressure. All of these technologies remain to be perfected. The long-term future of the coal industry seems to be rosy even though coal mining and combustion contribute a great deal to air pollution and global

HEAT ENERGY

Heat is one of seven different forms of energy on Earth: heat, chemical, mechanical, electrical, atomic, light, and sound. The amount of motion (usually referred to as activity) of atoms and molecules determines the amount of energy stored as heat, or *thermal energy*. Fuels used by power plants or vehicles contain a set amount of potential heat energy, which scientists can measure using a device called a calorimeter. A calorimeter burns a specific amount of fuel, and the heat given off by the burning then transfers to water flowing through the calorimeter. The difference in water temperature from when it enters the device and when it exits expresses the potential heat energy of the fuel.

fuel energy per volume = exit water temperature - entry water temperature

Industrial processes rely on various fuels for producing the heat they need; the heat energy is often captured and used in the form of steam. Common industrial fuels are coal, natural gas, oil, *coal gasification,* and solar. It takes energy to make energy, meaning that fuels consume a certain amount of energy in order to make them usable for industry. Coal requires the energy of large machinery to dig it out of the ground, and solar cells require energy to gather components and assemble the devices. At each step in these processes, a bit of heat energy is lost according to the second law of thermodynamics. For that reason industrial fuels are evaluated by a ratio called the *net energy ratio*, shown here:

usable energy ÷ energy used to produce the fuel = net energy ratio

warming. Problems associated with coal-fired plants, even plants using the best emissions-control technologies, suggest that clean manufacturing will need a major overhaul rather than small changes to reduce its wastes—the main objective of sustainable manufacturing.

POLLUTION CONTROL IN MANUFACTURING

Pollution can be controlled during manufacturing by using technologies that filter materials out of wastewater, collect emissions from smokestacks, or burn solid wastes. To control these types of pollution in the future, product teams meet to discuss a process they call "designing out

A ratio greater than 1.0 indicates that a fuel provides energy rather than consumes energy. The higher the ratio number, the higher amount of net energy a fuel produces.

Industries that need heat energy have tended to rely on surface- and underground-mined coals because these coals have net energy ratios of more than 25. By contrast, natural gas and oil have ratios of about 5, and coal gasification and solar devices have ratios of only 1.5 and about 1, respectively. When industrialists speak of the need for solar energy to "come down" before being profitable for business, they are talking about the net energy produced by solar devices compared with coal.

Even manufacturing plants that wish to rely on coal's heat energy can do some things to reduce wasted heat and thereby increase efficiency. The Energy Research Center of the Netherlands has described three technologies that may soon make traditional manufacturing more energy sustainable:

1. reheating waste heat so that it returns to a high temperature useful for the production process
2. storing waste heat for other uses, such as heating offices
3. transporting waste heat to use in other facilities or homes

Unless these or other technologies to manage heat energy become available at a reasonable price, traditional industries may have little incentive to dismantle their coal-fired power plants for cleaner energy-generating methods.

the waste." The team made up of product developers and manufacturing engineers develops new ways of making a product so that wastes are minimized at three points in the production: pre-production, in-process production, and customer supply. Some of the areas that manufacturers focus on to design out wastes are the following:

1. efficiency in raw material ordering so that only the amount needed is ordered

2. adjusting raw material use-up rate so that raw materials do not pass their expiration date

3. design of assembly line that reduces waste in the form of spills, leaking, or dust

4. collection bins that capture wasted raw materials and return them to the production process

5. reduced final product packaging

6. conversion to biological systems rather than chemical systems to reduce dangerous emissions

Steps such as these can be difficult to accomplish, expensive, or both. Since pollution control in manufacturing requires that profits do not decline, pollution control technologies must fit into a company's present operations. For instance, it does not make sense to install a smokestack scrubber on a plant that emits only steam but discharges large amounts of liquid chemicals. The scrubber represents an unnecessary cost and does nothing to address the pollution. Even government regulations, including legal actions and fines, have often been ineffective. The United States and many other countries have manufacturing plants that do not adhere to antipollution laws.

The World Bank has stated that tying economics to pollution control is the only way to clean up manufacturing wastes: "The new approaches are working because they have a solid economic foundation. Cost-minimizing plant managers will generally tolerate emissions up to the point where the expected penalty for pollution becomes greater than the cost of controlling emissions." This statement implies that industry will clean up its pollution only if forced to under the threat of severe penalty. Governments must therefore play a part in pushing manufacturers toward cleaner ways of doing things. As an example of this need, rapidly growing economies such as China's have made business advances faster than they have adopted pollution controls. The China State Environmental Protection Agency estimates that factory pollution accounts for more than 70 percent of the national total, 70 percent of organic compound water pollution, 72 percent sulfur dioxides in the atmosphere, and 75 percent of the dusts that carry particulate matter that damages the respiratory system.

New approaches to pollution control suggested by the World Bank must fit the industry. For example, power plants put emissions into the air, but the food industry is a leading producer of water pollution. The World Bank published a list in order of the top industrial sources of organic compound water pollution in its 2009 publication *Greening Industry*:

1. food
2. pulp and paper
3. chemicals
4. textiles
5. wood products
6. metal products
7. metal production
8. nonmetallic minerals

Many of the industries mentioned here have already started on new pollution-control programs. The table on page 98 gives examples of the many new approaches companies have taken to reduce the pollution coming from their manufacturing plants.

ZERO DISCHARGE MANUFACTURING

Zero discharge manufacturing encompasses all the methods that various industries use to eliminate the wastes—at least all hazardous wastes—they produce. Though each industry differs on how it does this, zero discharge manufacturing always has these things in common:

1. wastewater solid-liquid separation and reuse
2. minimized rinsing or cooling operations to reduce water use
3. nonhazardous solid waste cleanup and reuse
4. hazardous chemical elimination

EXAMPLES OF POLLUTION-CONTROL METHODS

NAME OF TECHNOLOGY	DESCRIPTION	INDUSTRIES
alternative blowing agents	replacing CO_2 for methylene chloride gas for expanding foams	foams, insulation, packaging
cleaning systems	replace xylene solvent with water and low-hazardous detergents	assembly lines
counterflow water system	reuses water and refreshes wastewater used in production process	metal plating
electric induction furnace	efficient waste-burner that minimizes the volume of chemicals and solid wastes	foundry
fibril	tubes produced by nanotechnology that replace plastics to reduce organic wastes	automotive, metal products, plastics
non- or low-HAP sealants and adhesives (HAP is hazardous air pollutants)	replaces solvent-based glues to reduce organic chemical emissions	automotive, furniture, packaging
water-based paints and coatings	substitution of water-based for oil-based products for equipment reduces organic chemical emissions	automotive, heavy equipment, metals

5. advanced filtration systems to collect hazardous organic compounds or metals

6. heat energy capture and reuse

7. scrubber, afterburner, or electrostatic precipitator for gas emissions

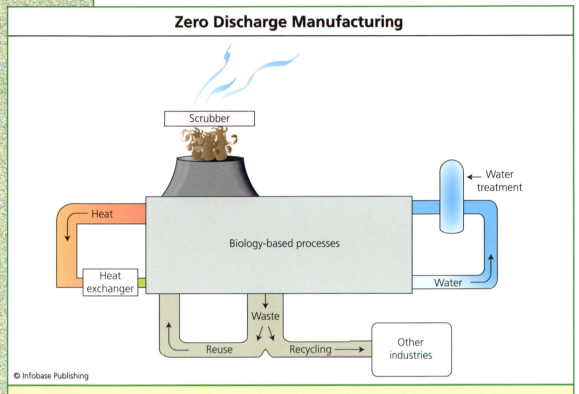

Zero Discharge Manufacturing

Zero discharge manufacturing is a method of making the industrial process much more sustainable than current practices. Zero discharge procedures usually center on biological methods of production that eliminate hazardous emissions and contaminated soil and water. Zero discharge emphasizes the reuse of materials, energy, and heat.

Even the dairy industry has attempted zero discharge operations, which represents an important breakthrough because agriculture of almost all types is a major producer of wastes. Zero discharge dairy farms may become more common as more farms adapt to the following practices: (1) capturing wastewaters and mixing with manure wastes; (2) collecting manure and treating in energy-producing on-site digesters; (3) directing rain runoff away from animal areas; (4) fixing all waste transport and feed storage leaks; and (5) building a secondary containment berm around animal and waste areas to hold wastes from accidental spills. Zero discharge agriculture is very difficult to achieve, but, if agriculture can make important headway toward sustainable methods, it will provide an important example for other industries to follow.

Current zero discharge operations usually begin with wastewater management for two reasons: water is a natural resource that is becoming increasingly scarce, and water carries waste chemicals produced by many different types of manufacturing. Zero discharge facilities have now incorporated one or more of the technologies described in the following table for controlling water pollution.

Zero discharge has been achieved in wastewater recovery and certain aspects within manufacturing such as production wastes, chemical wastes, and emissions. Manufacturers have found it difficult, however, to incorporate all three methods within one manufacturing plant. Any zero discharge method must consume less energy than the previous waste disposal method, or companies will be reluctant to accept it. The following "Case Study: The Energy Cost of Making a Car" explains why energy savings are vital for making businesses prosper.

WATER MANAGEMENT IN ZERO DISCHARGE MANUFACTURING	
TECHNOLOGY	DESCRIPTION
distillation	wastewater is boiled, releasing clean water as steam that can be condensed back into water
evaporators	devices accelerate the normal evaporation of water and collect the vapors that then condense back into water
reverse osmosis	pressure pushes water through a membrane filter, resulting in clean water on one side of the membrane and wastes remaining on the other side
ultrafiltration	water passes through a small-pore filter, resulting in clean water on one side of the membrane and wastes remaining on the other side
vegetation filters	plant life such as willows draw chemicals out of the water through their root systems, and harvesting the willows to use as an energy source

CASE STUDY: THE ENERGY COST OF MAKING A CAR

The life cycle of a car, which is the time between the car's assembly and the end of its useful life, draws upon dozens of industries and several energy sources to supply raw materials. The actual building of a car—stamping, body welding, assembly, and painting—uses less energy than all of the premanufacture steps. Carmakers have saved costs in body building by incorporating large amounts of aluminum since aluminum is a light metal that helps increase fuel efficiency and aluminum production demands less energy and so less money than other metals, such as steel. Calculations of the energy demand of a single car tend to consider the individual materials that go into the vehicle. According to this method, the following common car materials have varying energy requirements listed here in the order of their energy demand: wrought aluminum, copper, zinc, rubber, steel, and reinforced plastic. Wrought aluminum consists of metal made by shaping solid aluminum sheets; cast aluminum is made by pouring molten aluminum into molds.

How can wrought aluminum save energy in car manufacture if it demands more total energy to make? The answer comes from the significant recycling that the automobile industry uses. A single family car represents one of the most recycled products a person will ever buy. Aluminum holds an advantage over steel because, although it requires about 105 *megajoules* (MJ) of energy per pound (231 MJ/kg) to make, it costs only 24 MJ per pound (52 MJ/kg) to recycle. Steel, by contrast, requires almost as much energy to recycle (24 MJ per pound [52 MJ/kg]) as it does to make (30 per pound [66 MJ/kg]).

The total energy costs of building a vehicle depend on the size of the vehicle, number and type of components, and the current price for raw materials and labor. If carmakers design their future manufacturing plants to reduce coal as an energy source and turn to renewable sources, such as solar, wind, or water, the energy costs decline. The environmental group Green Car Congress has reported that manufacturers have been consistently cutting the energy costs of making a car; total energy costs

(continues)

(continued)

by manufacturers in Britain have dropped by one-half since 2001. Car production nevertheless drains resources and energy even with its extensive recycling. As a reference, the bicycle advocate Bicycle Universe has estimated it takes the same amount of energy to make 100 bicycles as it does to make one car.

Experts in the automotive industry have used various techniques to calculate the energy costs of making a car, but the calculations can be difficult to verify. Among all industries, vehicle manufacturing requires much less energy than petroleum and coal products, chemicals, food, paper, and iron, steel, and other metals. Vehicle production nevertheless requires energy for making the vehicle's raw materials—aluminum, steel, plastics, rubber, glass, fluids, forgings, textiles—and energy to run the assembly process. Accounting for all these factors, the United Nations Educational, Scientific and Cultural Organization (UNESCO) has calculated that a 1.1-ton (1 metric ton) car requires 25,600 MJ to produce—25,600 million *joules*. (One joule is the energy needed to lift an object weighing one Newton [0.445 pounds; 0.2 kg] 3.3 feet [1 m].) To put a single car's manufacture in perspective, the energy to make a car is about the same as the electrical energy needed to power an average house for one year.

SUSTAINABILITY AND BUSINESS

The Washington State University department of ecology has explained sustainability as "meeting the needs of the present without compromising the ability of future generations to meet their own needs." This goal could as easily apply to businesses as it does to the environment. Any activity that consumes resources so fast that it cannot sustain itself has created a trap that becomes difficult to escape. Today's large corporate business models ignore this basic concept by planning constant growth into the future even though the world's resources will not last forever. For businesses to be truly sustainable so that future generations can benefit from their products, business leaders must develop a long-term plan for the environment the same as they do for future profits.

In the mid-1800s in Butte, Montana, a generation of industrialists grew wealthy by dominating the silver, gold, and copper resources in the region. These copper barons, William Clarke, Marcus Daly, and F. Augustus Heinze, became the most powerful people for as far as one could see from Butte's town limits. But the metals did not come out of the ground forever, and by the middle of the 1900s the copper barons' fortunes and Butte's future plummeted. How difficult is it to draw a parallel between Montana's mining industry and today's oil industry? Though the Energy Information Administration (EIA) has predicted that new technologies will find another 76 billion barrels of oil in the United States by 2025, no one knows for certain the volume of oil still available. That is because the science of locating and measuring the size of as-yet undiscovered oil reserves contains a wide margin of error. Scientists do know that regardless of the volume underground, the oil will someday run out. Global warming that results from burning fossil fuels such as oil may well choke the planet long before the oil disappears. It is therefore in everyone's best interests to adapt to sustainable practices as soon as possible.

The most daunting challenge for businesses' conversion to more environmentally sound decisions comes from the business community itself. Bjorn Stigson, president of the World Business Council for Sustainable Development, remarked in 2008, "They [business leaders] know they cannot solve these problems alone, but have to work with others to develop solutions, even when this means learning to listen to their critics and those who oppose their actions." Each business's customers might hold the greatest power in getting industry leaders to listen.

Perhaps slow, steady improvements in business might give communities the best chance of success in converting to sustainability. Drastic changes often present big risks for business, but smaller steps toward sustainable practices balance environmental needs with business needs. For example, many companies have already reconfigured their activities to save on raw materials, reduce waste, conserve water, and conserve energy. These small steps have proven to be easy to implement and have a big impact over time. New business methods such as just-in-time production and more efficient distribution chains already help build profits while offering benefits to the environment. Meanwhile, companies have been expected to follow laws on emissions, waste discharges, and hazardous waste management and reporting. The next phase of decisions may incorporate some of these innovations:

1. conversion from coal-fired power plants to renewable energy sources such as solar

2. use of only alternative fuel for shipments

3. participation in waste recycling programs or transfer of waste to other industries

4. new technologies for handling water that cools production machinery and for returning the water safely to the environment

5. waste-to-energy processes

6. conversion from chemical synthesis methods to biological methods

7. redesign of packaging to reduce waste

8. redesign of products to biodegradable materials

9. conversion of offices to use of recycled products and alternative materials and energy

In business, the decisions that will make these innovations possible are referred to as *front-end* decisions, because they must be planned before the production process is designed, built, and operated. *End-of-the-line* activities, by contrast, refer to activities that try to make things better after all the production has been completed: disposing of wastes, installing devices to clean gaseous emissions, filtering discharge water to remove most of the pollutants, and taking back unsold products so they do not end up in landfills. The success of sustainability in business revolves around eliminating end-of-the-line activities by putting more emphasis on front-end decisions. Today, almost all major U.S. cities have sustainable business networks that help businesses implement sustainable practices. At this point in history, industries large and small have no excuse for sidestepping sustainable practices.

CONCLUSION

Sustainable manufacturing seeks to maintain a healthy business while operating in a way that does not harm the environment. Sustainable manufacturing therefore represents a major change from current business practices that have often put financial growth ahead of environmental issues.

Two major goals in sustainable manufacturing to protect the environment are waste and emissions reduction. The best achievement in this area is zero discharge manufacturing in which a production plant produces no hazardous wastes at all because all the hazards have been reused, recycled, transferred to other industries, or treated to make them safe. In order to achieve zero discharge or merely cleaner ways of making products, plants that have been powered by burning coal must finds ways to either make coal burning much cleaner or find alternative energy sources.

Sustainable manufacturing is in a sense a type of energy management designed to waste as little energy as possible by capturing all potential excess energy and reusing it in a manufacturing process. This theory requires a different viewpoint by industry, which has historically put profits ahead of environmental health. Technologies exist, and many more are on the way, to make manufacturing much friendlier to the environment. It may become the responsibility of environmental engineers, the public, and environmentalists to encourage industry leaders to move in the direction of sustainability. Sustainable manufacturing seems difficult to achieve, but it is not impossible.

5

ENERGY-EFFICIENT ELECTRONICS

Energy efficiency means the use of as little energy as possible to receive a benefit. For example, a kayaker knows to deliver even, smooth strokes to guide the kayak through the water rather than making choppy and exaggerated strokes, which power the kayak but waste energy. Both types of paddling take the kayak to shore, but the first method conserves energy so that the kayaker has stamina left over. Any piece of equipment can be designed with the same goal in mind: to perform a job with a minimum amount of energy wasted.

A switch to sustainable business begins by identifying all energy-wasteful activities and replacing them with energy-efficient activities. Office workers and homeowners receive help in doing this from energy utility companies that provide information on how to save electricity, natural gas, or heating oil. Today, offices and homes contain seven main areas of design that aid energy efficiency: (1) overall structure design; (2) landscaping; (3) electricity use; (4) insulation and sealing; (5) lighting; (6) heating and cooling; and (7) water heating. This chapter discusses electricity use and the electronic products that supply lighting, room heating and cooling, and water heating.

This chapters examines the special role electronic products, here called simply electronics, play in sustainable living. Today electrical engineers develop electronics that meet people's needs while also reducing overall energy demand on the environment. The chapter begins with the history of how society harnessed power for its various inventions through the centuries. The chapter then focuses on areas that are of special interest today:

solar energy for homes; energy-efficient appliances; lighting; heating; and communications. It also contains sidebars on the attributes of light as an energy source and how people measure energy. The chapter builds on the main theme of this book by explaining how engineers develop new technologies by learning from nature. Finally, it explores novel ideas in energy conservation: the use of sensors and feedback systems and the role of nanotechnology in energy production.

ENERGY EFFICIENCY THROUGH THE YEARS

The Sun has supplied the Earth with a never-ending source of energy since time began. As human life developed, the Sun remained the only energy source for human use unless a lightning strike happened to cause fire. The earliest humans probably ran from fires with other animals, but at some point in history they began to appreciate the heat that came from it. Fires caused by lightning strikes ran their course, burned out, and left humans again in darkness and cold, waiting for the next strike to rekindle the phenomenon. Perhaps humans learned to anticipate the next strike with the knowledge that light and warmth would return with it.

Humans' ability to start fires on their own represented a tremendous step in civilization because people had harnessed energy for the first time. Early humans started fire by rubbing two pieces of flint together, and they kept the fire going with the easiest fuel at hand—wood. A constant and controlled fire enabled people to live a less nomadic life, and small settlements congregated around fires for heat, light, and cooking. Meanwhile, residents of these settlements used the energy supplied by flowing water in streams and rivers to carry away wastes.

After mastering fire and making use of water's energy, Phoenician sailors 3,500 years ago learned to harness the wind to power their ships. Two thousand years later people combined waterpower and wind power to develop waterwheels and windmills to grind grain and run sawmills. As the centuries unfolded, oils that bubbled up from the sea, coal, and natural gas all served as fuels for heat and making light. Though these fuels dominate the industrialized world today, until about 150 years ago wood remained the main source of energy, as it still is in many underdeveloped parts of the world. (Highlights in the history of energy development are summarized in Appendix C.)

Throughout the 20th century and the early years of the 21st, oil, coal, and natural gas served as society's leading sources of energy. But prolonged use of these fossil fuels has led to water and air pollution, destruction of habitat, and global warming. The 21st century will make greater use of six energy sources that avoid the use of fossil fuels: solar, wind, wave and tidal action, geothermal, hydrogen, and *biomass*. Of these energy sources, solar energy has provided the first breakthroughs in shifting a fossil fuel–based society to one that depends on nonfossil fuels. The California sustainability expert Geof Syphers said of the new solar communities on the rise, "This is really about lifestyle. We want to make sustainable living easy for people. We want to make it appealing." Like fire, wind, and water harnessed by the ancients, steam engines and combustion in the Industrial Revolution, and modern nuclear energy, new energy sources are always appealing because they enable technology to take a major step forward.

Solar communities like that described by Syphers and other renewable energy sources have been supplying an increasing portion of total energy consumption in the United States, but these alternative energies still account for only 7 percent of total energy consumption. (Petroleum supplies 40 percent, natural gas supplies 23 percent, coal accounts for 22 percent, and nuclear energy supplies the remaining 8 percent of total U.S. energy consumption.)

The solar power industry divides its business into two segments: photovoltaic power, which converts sunlight to electricity, and thermal solar power, which turns sunlight into heat. Photovoltaic systems have grown rapidly in the industry for the past 10 to 15 years, and they make up a large portion of renewable energy's 3 percent annual increase in growth. Thermal solar power rose for the 10 years up to 2006 as oil prices increased, then declined as oil prices declined. The outlook for the global solar industry continues to be encouraging, especially because many new solar technologies have been emerging for the purpose of increasing solar power's efficiency and decreasing its cost.

SOLAR HOMES

Solar power for homes and other buildings uses either an active or a passive form. Active solar devices collect energy from the Sun and store it or move it by the use of electrical controls, pumps, and fans. Passive solar power, by contrast, does not rely on any mechanical help. Passive solar

Houses with solar panels such as this one drastically cut the electricity they take from the municipal power grid. *(Gray Watson and Rosemary McCrudden)*

energy employs windows, walls, and doors to collect, store, and distribute solar energy without the need for any other energy input. Passive systems and active systems conserve heat in the winter, but they also reject excess solar heat in the summer.

The main principle of solar energy relates to the concept of *gain,* which is the amount of usable solar energy that a structure captures. Solar homes can be designed to contain one of the following three types of gain:

1. direct gain—sunlight enters south-facing windows and strikes walls and floors, which store the energy as heat
2. indirect gain—solar heat strikes the home's outside wall and is stored between the outer and inner wall from which it radiates as heat

3. isolated gain—a structure, such as a solarium, separate from the main house captures and stores solar energy as heat in its masonry

Regardless of the type of solar gain a house has been built to use, designers and environmental engineers pay attention to five basic elements of solar energy systems, described in the following table.

Solar homes distribute energy by three different means: conduction, convection, and radiation. In conduction, heat energy moves through solid matter by exciting molecules as it spreads through the matter. For example, a lit stove boils water in a pot because the metal pot conducts the stove's heat energy to the water. Convection consists of heat circulation through

Solar energy can be equally valuable outside traditional neighborhoods, such as this passive solar farmhouse in New York. *(Green Rabbit Farm)*

FIVE ELEMENTS OF PASSIVE SOLAR HOMES	
ELEMENT	DESCRIPTION
aperture	large windows through which sunlight enters
absorber	hard, dark surface situated in the direct path of sunlight that stores the energy as heat
control	any device that reduces sunlight exposure in summer and increases exposure in winter
distribution	devices for distributing the stored solar energy throughout the house
thermal mass	any surface outside the direct path of sunlight that stores the energy as heat
Source: U.S. Department of Energy	

liquids or gases rather than solids. A lighted fireplace, for instance, heats a room by convection. Radiation may be solar or infrared radiation, both associated with the light spectrum, or nuclear radiation used in nuclear power plants. Solar radiation consists of the movement of heat through air from warm objects to cooler objects, such as a solarium warming up when exposed to bright sunlight. Infrared radiation consists of a warm surface transferring heat to a cooler surface. This type of radiation works when a person is chilly and snuggles under an electric blanket to warm up.

Solar homes contain design elements and building materials that aid in heat distribution by any of three different mechanisms: radiation is absorbed, reflected, or transmitted. Large glass windows transmit plenty of sunlight to the indoors. Glass absorbs only 10–20 percent of solar radiation and transmits the rest. The 80–90 percent of solar radiation that enters a house can then be absorbed by materials that absorb and hold heat better than others. Opaque solid materials such as stone floors and walls absorb 40–90 percent of the radiation entering a house; darker materials absorb more heat than lighter colored or white materials that reflect much of the radiation.

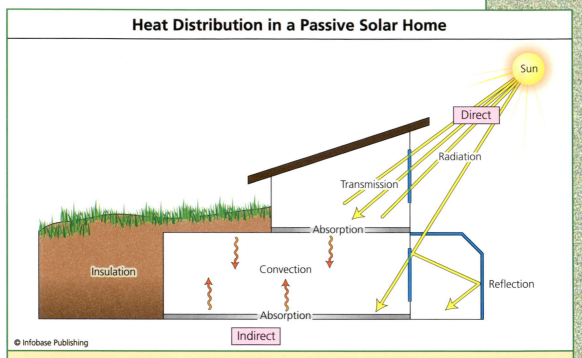

Heat Distribution in a Passive Solar Home

Passive solar energy works best if a building has been designed to make maximum use of the sunlight for heating and energy production. This diagram illustrates the ways in which passive solar energy collection is optimized—large windows, reflective surfaces—and stored in thermal mass materials, such as concrete and brick, which absorb heat then gradually release it.

The best-designed solar homes possess an attribute called *thermal capacitance,* meaning the capacity of materials to store heat. The following materials predominate in new solar homes because they have a high degree of thermal capacitance: stone, masonry, concrete, brick, tile, and water. These materials are named thermal mass materials and may also be thought of as insulating materials because they hold their temperature once they have heated up or cooled down. Thermal mass is so effective that in cities where hundreds of brick and concrete buildings fill each block, the buildings raise the outside temperature 2–8°F (1.1–4.4°C), an event known as the *heat island effect.*

Many solar homes offer all the luxuries of conventional homes (modern fully equipped kitchens, storage space, patios, porches, pools, fireplaces, skylights, fountains, and gardens), and they have become the most desirable type of new homes in many parts of the United States.

The energy expert Bernadette del Chiaro of the alternative energy advocacy group Environment California told the *Los Angeles Times* in 2007, "[House] builders are seeing that they'll get more buyers coming to their developments when they have solar. They sell like hot cakes." Solar homes in fact outsell homes in some communities such as San Jose, California. Solar homes nonetheless account for only a little more than 5 percent of all homes in California, a state that seeks to reach 50 percent solar in all new housing by 2015. Other smaller communities have similar goals of making all new public buildings solar-powered.

Solar devices installed on homes work best to save energy when used in combination with thermal mass building materials, as mentioned. Other alternative building materials sell at moderate cost, but the use of these materials requires a break from traditional methods of construction. Michael Funk helped design his solar-powered house in California's Sierra Nevada. He admits that not all alternative materials were as readily available at building supply businesses as conventional materials. "When it comes to building with alternative and sustainable materials and using local artists and craftspeople, you can run into resistance," he told *Natural Home and Garden* magazine in 2006. "You have to keep asking, keep digging, keep pushing. The products are out there. When you meet resistance, just push back. There's a way to get everything you want. It just takes more work and persistence." The table on page 114 lists common components and materials preferred for building new solar homes.

Existing houses and buildings can be switched to solar power with the installation of solar panels on the roof and connections to the house's main energy input. The task can be done in a few days to weeks depending on the size of the building and number of panels to be installed. Many homeowners have been hesitant to install solar energy because of solar's current high price. For the present, homeowners need about 50 years for the energy savings of solar energy to repay the installation cost.

Homeowners cope with three disadvantages to solar power in addition to expense: (1) periodic cutting of overhanging tree branches that block the sunlight from reaching rooftop collectors; (2) surrounding tall buildings that may make solar energy use difficult or impossible; and (3) solar collectors on the house that change the house's appearance. Solar homes in very cold climates or places with heavy cloud cover may need a backup generator or occasionally draw energy from

COMPONENTS OF SOLAR HOMES

COMPONENT	ATTRIBUTES
roofing	
lead-free aluminum or steel, coated ceramic, recycled asphalt shingle, clay tile, slate, fiber-cement composite, recycled plastic-rubber shingle	durable, fireproof, recyclable, lighter materials reflect heat
structural frame	
Forest Stewardship Council certified wood, reclaimed lumber, engineered lumber (composites made from sustainable, fast-growing wood)	conserve old-growth forests, avoid glues that emit toxic gases
insulation	
expanded polystyrene foam, fiberglass, straw bales or other cellulose sources, recycled textiles (example: jeans)	avoid release of organic gases, reduces the use of ozone-depleting materials
electricity	
active solar collectors, photovoltaic panels	spares use of community electricity, reduces dependence on coal-fired power plants
heating, ventilation, air-conditioning (HVAC)	
heat pump or heat exchanger, EPA-approved wood burning stove, natural ventilation with windows, ductless air-conditioning to specific rooms	reduces overall electricity demand
water heating	
tankless flash heater, heat pump water heater, insulated small tanks	reduces overall energy demand, conserves water

the local municipal supply. Solar homes also offer many advantages that outweigh some of the concerns. First, they use an energy source that is free, sunlight. Second, they produce no carbon dioxide (CO_2) emissions, and, third, produce no or little air, water, or noise pollution. Fourth, rooftop solar collectors install quickly, and, fifth, collectors do not upset the landscape. Finally, the energy cost savings drop to near zero or even below zero. In some U.S. states, energy utilities reimburse homeowners with solar systems that put energy back onto the community's *energy grid,* also called a power grid.

Solar power has become mainstream throughout Europe, Australia, and the United States because it uses the simple principle of letting nature supply energy for human use. Ecosystems have always managed energy this way, but human civilization long ago adopted a different means of making energy: producing it by burning energy-containing fuels in engines. Environmental engineers are beginning to return to studying nature for clues about how to run systems at maximum efficiency from natural materials, the principle behind biomimicry. The "Case Study: Learning from Electric Eels" on page 116 examines a unique energy-producing model in nature.

SMART APPLIANCES

The core idea of sustainable living pertains to energy generation, use, or waste. Environmental engineers have investigated opportunities for energy saving inside homes and other buildings. One of the most familiar opportunities comes from Energy Star appliances that have been designed and certified to ensure they conserve energy compared with older appliances. The next step in energy savings will be in the emergence of smart appliances, which consist of home appliances, such as refrigerators or washing machines, equipped with computer chips that sense energy usage and regulate the energy consumption of the appliance. For example, a smart refrigerator may contain a chip that tells the compressor to go into a rest mode when the unit's light (which indicates an open door) stays on for longer than a minute. This action prevents electricity waste during the refrigerator's use.

Other appliances can employ similar rest modes in which the appliance partially shuts off to save energy. Freezers, washers, dryers, water heaters, dishwashers, ranges, and microwave ovens would be helped by this technology. Most homes in the United States have all of these appliances,

This unique smart appliance combines a toilet with a washing machine. The wastewater from the washing machine goes to the toilet tank for flushing. *(Peazyshop.com)*

and most of them run at the same time. By adding up all the homes on a single block, then totaling the households in a moderate-sized town, it takes little imagination to guess the effect these appliances have on a community's power grid when they all run concurrently.

The next generation of home-based smart appliances will probably connect by Internet to the municipal power grid to reduce power drains during the heaviest usage periods. This type of appliance–power grid connection provides two advantages: overall energy conservation and alleviation of demand on grid infrastructures during peak times when they are most vulnerable to breakdown. These same appliances may also sense when the power grid is at its most stressed and automatically go into rest mode. A household filled with smart appliances would balance the community's power use and waste automatically for the homeowner.

Electronics in addition to large kitchen and laundry room appliances have also been planned for automatic energy regulation. Televisions, video equipment, home entertainment systems, telephones, and computers all draw energy even when they are turned off. Many people have learned to unplug these devices when not in use, but new smart electronics will make this unnecessary and be more dependable in conserving energy.

Smart appliances resemble any innovation in technology in that their costs will be high at first and then become more affordable as the public purchases them. Rob Pratt, a program manager at an appliance testing laboratory, remarked to MSNBC in 2007, "If this becomes cheap enough, even your coffeemaker can help the grid out." The MSNBC report concluded that if the majority of homes in the United States converted to smart appliances, the country would save about $70 billion in new power plant construction and power distribution costs in 20 years. By turning energy-saving control over to appliances, people may receive an overall benefit in energy and money savings.

LIGHTING

People spend most of their lives indoors in industrialized countries, so lighting becomes an important drain on a power grid. Lighting comes in two forms, natural sunlight and electrical, and each of these can be managed in a way that increases the overall energy efficiency of a home.

The most efficient use of natural sunlight is called *daylighting*. Daylighting integrates natural light with electrical light to create the best indoor lighting conditions as well as to save energy. Daylighting usually involves the following four main structures for maximizing the amount of sunlight that enters a structure:

1. light pipes—roof to ceiling tubes, about 13 inches (33 cm) in diameter, that can light about 200 square feet (19 m²) of space with sunlight

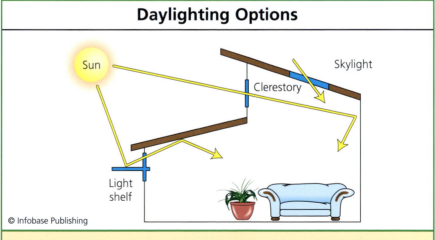

Daylighting Options

© Infobase Publishing

Daylighting is an energy-saving technique that reduces the need for electric lights during the day. Many techniques aid daylighting. A clerestory or a light shelf each reflects light deeper into the house's interior. Special glazes on wall surfaces also help reflect light into rooms.

2. skylights—large, rectangular rooftop windows

3. clerestory windows—narrow horizontal windows set high into walls

4. light shelves—horizontal platforms that redirect light hitting a large south-facing window so that the light penetrates deeper into a room

The daylighting structures mentioned here provide benefits to the home in addition to bringing in more light. For instance, clerestories opened in summer let hot air out, and if they are made of well-insulated glass they hold in warm air in the winter. Light shelves expand the available light to make electrical lighting almost unnecessary during the day, and they reduce glare. The Lawrence Berkeley National Laboratory has

CASE STUDY: LEARNING FROM ELECTRIC EELS

Biomimicry may be considered a science of emulation in which engineers build systems for human use based on the way nature builds its systems. This science looks at nature in a new way, focusing not on what humans can extract from nature but rather what they can learn from nature.

In 2008 researchers at Yale University developed tiny artificial cells for the purpose of powering medical implant devices. They used the energy-generating cells called electrocytes in electric eels as their model. The chemical engineer Jian Xu explained, "The electric eel is very efficient at generating electricity. It can generate more electricity than a lot of electrical devices." (A clue to this ability lies in the eel's biological name, *Electrophorus electricus*.) The engineering team's problems in generating energy the same way as an eel does were twofold: understanding how the electrocyte works and learning to build a similar electrical device.

Electric eels have three different energy-generating systems: two high-voltage systems are used for defense and for stunning prey and a low-voltage system helps in navigation. Disc-shaped electrocytes run all three of these systems. The electrocytes stack up like a series of watch batteries in the eel's organs so that when each fires and produces a low to moderate amount of energy, the cumulative effect is a large energy pulse.

Non-firing electrocytes hold a negative charge inside the cell by constantly pumping positive sodium ions (Na^+) out and allowing positive potassium ions (K^+) to naturally diffuse out through

estimated that light shelves can expand the indoor daylight zone 2.5 times the window height.

Improved morale and even health may be added benefits that come from daylighting. Students in classrooms and office workers have better attention, memory recall, and productivity in daylighted rooms than people in rooms with only artificial light. Building designer David Hobstetter wrote in a well-researched 2007 article in *Real Estate News,* "Our lives have become increasingly sedentary and cloistered over the past few decades, with the rise of the digital age. As a species, we now pass the vast majority of our time indoors . . ." Hobstetter attributed increased stress, fatigue, absenteeism, and even work-related illnesses to poor indoor working conditions, including a lack of natural light. A Toyota employee put it in simpler terms when talking with the *Los Angeles Times* reporter Roger Vincent, "The [natural] lighting is easier on the eyes and on the nerves. It's

the cell membrane. An ion is an element missing electrons or possessing extra electrons. To fire the electroctyes the eel's brain sends a message to nerve cells, which stimulate one side of each electrocyte. The side of the electrocyte receiving this nerve impulse becomes stimulated in a process called *depolarization.* The nerve-side of the electrocyte becomes depolarized an instant before the far side of the electrocyte depolarizes. This occurrence leads to a temporary one-way flow of ions or an electrical charge. The eel's charge of up to 600 volts comes from the synchronized depolarization of about 200,000 electrocytes.

Yale's researchers built small discs based on the electrocyte's depolarization action, which they called a "bio-battery." The quarter-inch- (0.64-cm) thick bio-batteries contain the following two components: (1) artificial membranes based on the electrocyte membrane that sets up inside-outside charge differences, plus (2) proteins that mimic the ion channels in real membranes. So far, the artificial electrocytes have been made to generate 30–40 percent more power than the natural electrocyte. The researchers plan to line up the bio-batteries in several stacks of about a dozen to produce enough electricity to power medical prostheses such as retinal implants.

Though these plans and devices are scarcely off the drawing board, tiny power generators have opened similar possibilities for using bacterial cells or even mitochondria—the energy-generating component of eukaryotic cells—as mini–power plants. The task involves only the willingness to apply basic engineering concepts to designs produced by nature.

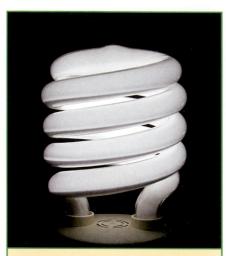

The compact fluorescent lightbulb demonstrates that not all achievements in environmental engineering need be complicated to have a positive effect on the environment. By redesigning normally long, straight fluorescent lighting into a compact size about the same as an incandescent bulb, engineers introduced an energy-saving product for homes and offices.

pleasant and feels more productive." Daylighting therefore has psychological advantages to complement health and energy-saving benefits. A description of natural light can be found in the following sidebar "Light" on page 122.

Electrical lighting serves as a substitute for natural sunlight and offers the following advantages over sunlight: (1) it can be turned off; (2) it can be dimmed or made to produce high intensities; (3) it can be selected in different colors; and (4) it provides light in places where sunlight cannot reach. The best way to improve energy efficiency in electrical lighting is to replace all standard incandescent lightbulbs with either compact fluorescent (CFL) lightbulbs or halogen lights. Incandescent bulbs emit only 10 percent of their energy as light and 90 percent as heat, meaning they are very inefficient at converting electrical energy into light energy. CFL bulbs cost more than incandescent bulbs, but they last 10 to 20 times longer and use 75 percent less energy. The following table summarizes the impact of CFL bulbs on energy savings by using the U.S. Environmental Protection Agency's (EPA) and Department of Energy's (DOE) joint Energy Star online calculator.

Each incandescent bulb replaced by a CFL bulb reduces the CO_2 emissions from power plants by several hundred pounds over the bulb's useful life. If each U.S. household replaced one light, CO_2 emissions would drop by more than 1 trillion pounds (454 billion kg). Halogen lighting does not save as much as CFL bulbs, but halogen lighting is still about 50 percent more efficient than incandescent lighting. All types of electrical lighting can be made more energy efficient by using light sensors to dim or turn off lights on bright days and timers to turn off lights automatically.

SAVINGS FROM USING CFL BULBS	
BENEFIT	SAVINGS COMPARED WITH INCANDESCENT
bulb life	$326
bulb life energy	2,100 kWh
bulb life air pollution reduction	3,224 pounds (1,462 kg) CO_2
savings as percent of retail price	782 percent
air pollution reduction (equivalent to vehicles)	0.28 car removed from roads per year
air pollution reduction (equivalent to forests)	0.4 acre (0.0016 km²) of forest saved per year

The U.S. Congress has voted to follow the lead of the European Union, Australia, New Zealand, Japan, and Brazil in banning the sale of incandescent lightbulbs beginning in 2012. According to the plan, the United States will have phased out all incandescent lightbulbs by 2014.

HOME ENERGY AND HEAT STORAGE

The United States wastes almost 85 percent of all the commercial energy it takes off the power grid. People waste by not conserving energy in their daily activities: leaving lights on, letting television play with no one watching, or running small laundry loads. Faulty appliances and inefficient machines and vehicles waste additional energy. Energy waste also occurs naturally due to the second law of thermodynamics that states that some energy is always lost when being converted from one form to another. Considering these opportunities for potentially usable energy to escape, engineers now design structures that have several different systems for conserving, recapturing, and reusing energy, whether the energy is in the form of light, electricity, or heat.

LIGHT

Scholars have puzzled for centuries over the nature and the meaning of light. Light has been thought of as streams of particles since the time of the ancient Greeks. Though no one had actually seen the light particles, scientists assumed the particles simply moved too fast for the human eye to detect, so humans perceived light as a beam or as a field of light. In the 1600s Christian Huygens described light as having more characteristics of waves than of particles. For the next 300 years leading to studies by Albert Einstein, scientists built upon this theory that light behaved like a wave.

Today physicists describe the complex concept of light either as the particle theory or the wave theory. The particle theory states that light consists of beams of photon particles, which Einstein defined as distinct energy packets. The wave theory, also clarified by Einstein, suggested that light is made up of light waves. In fact, light behaves as both particles and waves, but physics students tend to study light in terms of waves because it is easy to understand wave behavior and apply it to light's characteristics. Lightbulbs serve as a convenient source of light that follow all the rules listed here. The most familiar characteristics of light are the following:

1. Light reflects off surfaces at the same angle at which it strikes the surface.
2. Light waves do not need a medium such as water to travel through (sometimes this is expressed as the ability of light to travel through a vacuum).
3. Light waves radiate from their source, so light is referred to as radiation.
4. Light waves' energy is in the form of electric fields and magnetic fields, explaining why light can be described as electromagnetic radiation.

New homes contain many improvements in energy management that older houses from the 1950s through the 1980s did not possess. Insulation, seals, doors, windows, and roofs have all been improved for the purpose of retaining household energy, usually in the form of heat.

A new type of construction called a superinsulated house contains features so efficient in holding heat that sunlight, appliances, and residents provide enough heat to keep the indoors warm. Superinsulated houses contain some or all of the following features: (1) very thick insulation with R values of R-40 to R-60, meaning they have high *thermal*

5. Light occurs in a range of *wavelengths* that begin as invisible (to humans), progress into a visible range called the *visible spectrum*, and end again with invisible waves.

Light emanates from both incandescent bulbs and CFL bulbs with very little visible difference, but each light source works differently. Any light source produces light by energizing atoms, which causes the atoms' electrons to behave in an excited state. Electrons are negatively charged pieces of an atom. Eventually, the electrons fall back from their excited state to their normal state in the atom, and this transition from higher to lower energy releases extra energy in the form of a light photon. Incandescent bulbs energize atoms by heating them, which explains why these bulbs become very hot when they are turned on. Fluorescent bulbs use electricity rather than heat to energize atoms. In fluorescent bulbs, an electric current flows from one pole to an opposite pole. During this flow, the current's electrons bump into mercury (in vapor form) atoms along the path. The energized mercury atoms emit ultraviolet light as they return to a low-energy state; the ultraviolet light hits a compound called phosphor, which then emits visible light.

Environmental engineers treat light as other scientists do, that is, light is a form of energy and in this form the universe transmits energy from one place to another. Solar homes equipped with light collectors (called photovoltaic cells) convert light energy into other forms of energy, usually electric and heat. Meanwhile humans depend on light's interaction with their ocular system (the eyes and the nerves connecting them with the brain) to interpret their surroundings. Light has certainly been taken for granted by nonscientists in their everyday routines, but this abundant resource is becoming one of the most promising options for shifting away from dependence on fossil fuel energy. Light energy may soon become the main energy source for cars, boats, appliances, and telecommunications in addition to new solar homes.

resistance and so prevent heat transfer across their boundaries; (2) triple-paned insulated glass in windows; (3) few or small north-facing windows; (4) airtight seals; and (5) an air-to-air heat exchanger. Air-to-air heat exchangers are appliances that transfer heat across a barrier from warm, stale indoor air to cool, fresh outdoor air. By this process, heat exchangers ventilate the indoors and capture and reuse heat to prevent heat waste. Superinsulated houses are also likely to contain thermal mass materials that hold heat. (Appendix D describes devices used today for electrical energy storage.)

Older houses often waste energy as heat loss that can never be recaptured, but even these houses can be improved with simple fixes. The following steps help transform an energy-wasteful house into an energy-efficient house:

1. plug leaks around windows, doors, in cracks, and in holes
2. replace old insulation
3. replace old windows with insulated, energy-efficient windows
4. install a heat exchanger or a heat pump (removes warm air and replaces with cool air)
5. use energy-efficient appliances and lighting
6. replace large tank water heaters with a tankless instant water heater
7. install good seals on doors to attics and unheated basements and keep doors closed

American architect Edward Morse designed a wall in the late 1800s in which a sheet of glass covered the wall's masonry. This design increased heat transfer from outside to the inside of the building and reduced heat loss. In the 1960s French designer Félix Trombe improved on the concept, and the wall now bears his name. The Sun's heat passes through a doubled-paned window and warms the masonry next to it. The masonry then releases the heat into the interior. (*MultiwallSystems.com*)

Homeowners need not wonder how well they do in saving energy because each home has a meter that shows electricity and heating fuel usage. The basics of energy measurements used by energy utilities are described in the sidebar on page 126 "The BTU and the Kilowatt."

SENSORS AND FEEDBACK

Sensors that work with feedback systems easily and automatically conserve energy. Energy sensors monitor the use of electrical and heat energy within a room or an entire building. Sensors can be

used in two different ways to conserve energy. First, they monitor a
building's peak and nonpeak times
of energy use and so identify the
periods in which a structure puts
more or less stress on a community's power grid. Second, on-off
sensors such as motion sensors
turn indoor or outdoor lights on
when they detect motion and off
when the area is no longer being
used to reduce energy waste.

Sensors in combination
with feedback systems provide
even better control of energy use
because a feedback system tells a
building's residents how efficient
they are in using energy. A sen-
sor-feedback system contains two
components: the detection unit
called a transmitter that may be
attached to an electrical or gas
utility meter, a water meter, or a
thermometer, and the display unit
located inside the house, which

A wood stove generates heat without
using electricity. Stoves and fireplaces
conserve a house's solar power and avoid
the need to draw from the municipal
power grid. Wood is a type of biomass
that produces an equivalent amount
of Btu as crop wastes, which are also
combusted for energy production. *(Jotul)*

receives wireless signals from the transmitter. Sarah Rich wrote for the
online environmental magazine *WorldChanging,* "It's one of those telling
facts of human nature that when we are being monitored, our behavior
changes." Just like a speedometer in a car, people self-monitor when they
receive helpful information. "The same holds true in households," Rich
explained, "where inhabitants can be made immediately aware of their
energy consumption. If you can see your pennies piling up on account of
a light you left on in the bathroom, you can bet you'll remember to turn it
off." These so-called smart sensors offer one of the simplest ways to manage energy use.

New systems make energy savings even easier for people to save
energy. Hotels or homes can be equipped with key cards that contain
information on the indoor heating, cooling, and lighting systems. As a

THE BTU AND THE KILOWATT

The acronym Btu (or BTU) stands for British thermal unit. Energy utilities use two terms, Btu and kilowatt (kW) to describe energy and power; a Btu is a unit of energy and a kW is a unit of power. Energy equals the potential ability to do work or produce heat. By contrast, power is the rate at which work is done. The difference between energy units and power units can be expressed as follows:

1. A kilowatt is the rate of energy use at this instant.

2. Kilowatt-hours refers to the total amount of energy used over time.

Therefore, utility companies, appliance companies, and fuel producers often relate the Btu to the kilowatt by transforming the expression for energy to an expression for power by the following conversion:

$$1 \text{ Btu} = 2.9 \times 10^{-4} \text{ kilowatt-hours (kWh)}$$

(The Btu itself can be converted to a unit of power, the Btu-hour [Btu/h], though this term is less commonly used than kilowatt-hour.)

A person can visualize the Btu as the amount of heat required to raise one pound (0.45 kg) of water one degree Fahrenheit from 60°F to 61°F (15.6° to 16.1°C). Home appliances range from 5,000 to 50,000 Btu, and in the United States the population uses 100 quadrillion Btu (called *Quad*) in a single year.

Most home electrical devices use very small kilowatt amounts that can also be measured in watts, or one-thousandth of a kilowatt. Electrical devices typically range from 15 to 500 watts (0.015–0.5 kW). An appliance that provides a power value in amps, for amperes, rather than watts can be converted as follows:

$$\text{number of amps} \times 120 \text{ volts} = \text{number of watts}$$

By understanding the Btu and kilowatt-hours, homeowners can accomplish two objectives: gain a better idea of how appliances consume energy, and understand their rate of energy use in order to monitor energy savings.

person enters a room, inserting the card into a small slot box near the door, lights and other systems turn on to make the environment comfortable, with heat, air-conditioning, or ventilation. The person simply removes the card from the slot when they leave, and the systems readjust to energy-saving mode. Sensors and automatic on-off switches are inexpensive, and the electricity savings make up for the price in a short period of time.

ENERGY FROM NANOTECHNOLOGY

Energy-efficient electronics depend on some form of energy storage so that excess energy can be held for future use and not wasted. In fact power storage makes up one of the three main components of energy management for homes as well as vehicles: power generation, power transmission, and power storage. Lynn E. Foster, author of the book *Nanotechnology: Science, Innovation, and Opportunity,* opined, "If you can solve the problem of local storage, you've basically solved the whole problem. That's because, by definition, the storage you need is local. You've got terawatts [equals one trillion watts or one billion kilowatts] of power moving into the grid. The biggest problem with renewable energy in general, and solar and wind in particular, is that they're episodic and not dispatchable." This means that the energy supplied on community power grids comes in pulses rather than in a steady stream, and utilities cannot dispense the energy in doses as needed. Foster continued, "You've got to have storage. Storing energy in batteries, capacitors, fuel cells, and some chemical systems like hydrogen depends on nanoscale interactions." *Nanoscale* is a general term for the size of things that are no bigger than an atom or a molecule—a nanometer is one billionth of a meter. Nanotechnology is the science of working with nanoscale-sized devices.

Nanotechnology's value to science lies in the observation that matter behaves differently at the nanoscale than it does on a larger scale. The DOE has explained this technology and its value: "All the elementary steps of energy conversion (charge transfer, molecular rearrangement, chemical reactions, etc.) take place on the nanoscale. Thus, the development of new nanoscale materials, as well as the methods to characterize, manipulate and assemble them, creates an entirely new paradigm for developing new and revolutionary energy technologies." Nanotechnology is still in its

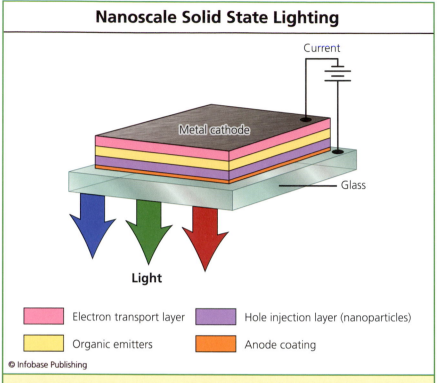

Nanoscale Solid State Lighting

Current

Metal cathode

Glass

Light

■	Electron transport layer	■	Hole injection layer (nanoparticles)
■	Organic emitters	■	Anode coating

© Infobase Publishing

Engineers have been using nanotechnology to develop thin layers designed for generating an electrical current and producing light. The next step using nanoscale materials may be to convert solar power directly to light with a solid-state structure such as this one. Engineers also plan to embed tiny solar cells in window glass to produce electricity from sunlight. *(Source: Paul Alivisatos, Lawrence Berkeley National Laboratory)*

infancy, as scientists discover new behaviors of common materials when at the nanoscale level and find ways of applying this knowledge for practical uses. The following table provides a list of the main opportunities in nanotechnology for producing new energy-generation systems for electronics and other items.

Nanotechnology will likely help in the development of the following two energy conservation objectives: inventing future energy generation devices that perform in ways that current devices cannot, and finding new ways to catalyze reactions to minimize energy input and waste output. At present, ideas for new uses of nanotechnology have developed faster than actual methods have developed.

NANOSCALE ENERGY TECHNOLOGIES		
TECHNOLOGY	**DESCRIPTION**	**POTENTIAL APPLICATIONS**
catalysis	controlled chemical reactions that avoid hazardous chemicals and high heat conditions	energy-efficient and clean manufacturing; production of new fuels
hydrogen production and hydrogen storage	water-splitting reactions	hydrogen fuel energy cells or nanotubes for vehicles, electronics
nano-materials	unique molecular structures	lightweight materials for vehicle bodies and electronic components
solid state lighting	layered structure containing a positive-charged metal, a color-producing polymer, and a negative-charged metal emits various colors when different polymers are activated	light sources at 50 percent of current power consumption
nano-crystals	familiar materials rearranged in crystalline structure for the purpose of transferring energy	conductors; heating elements
high efficiency solar cells	sunlight hits a material containing nanoscale pores, which facilitates charge transfer	energy sources for various devices
ultrahigh efficiency solar cells	nanoscale prisms break sunlight into individual sections of spectrum to create specific solar energy capture with no waste	energy sources for various devices

(continues)

NANOSCALE ENERGY TECHNOLOGIES *(continued)*		
TECHNOLOGY	**DESCRIPTION**	**POTENTIAL APPLICATIONS**
solar paints	restructured materials in crystal form made to transmit charge when exposed to sunlight	energy sources for various devices and vehicles
solar cell-LED devices	convertible layered devices that produce electricity when exposed to sunlight or light when powered by electricity	materials for homes, buildings, electronics
filters	materials with nanoscale pores to remove molecules of hazardous compounds	water treatment; medical solution sterilization; medical treatments at the blood-brain barrier
nano-sensors	devices imbedded in materials for detecting instantaneous changes in temperature, pressure, humidity	coatings for energy-efficient aircraft, ships, vehicles

CONCLUSION

Electricity furnishes a large portion of the world's energy demand for residences and commercial buildings, and most of this electricity goes to running small appliances, larger equipment, heating and cooling, and lighting. Environmental engineers have vast opportunities to help build sustainability by devising items that follow a new scheme for energy generation. Today's electricity generation comes mainly from coal-fired or natural gas-powered electric plants that put many quadrillion Btu of electricity on the power grid, which in turn supplies, houses, offices, and factories. Due to pollution from this method, plus the fact that these nonrenewable resources are limited, engineers can serve sustainability by designing new ways to generate electricity and new electronics that use less electricity. Environmental and electrical engineers have begun to make progress in both of these approaches to energy management.

People in industrialized nations have taken electricity for granted to the point where lighting, climate control, and instant means of communication, cooking, or entertainment are commonplace. Sustainable use of electricity will be accepted only if the new way of doing things provides the same benefits quickly, easily, and inexpensively. In a word, electronics that contribute to sustainability must be efficient.

People can use their electrical energy less wastefully by simply paying attention. Actions such as using daylighting, turning off appliances and lights when not in use, and monitoring use make a great impact on lowering total power consumption. These actions require no more than a modest change in behavior, and people have little excuse for ignoring the need for reducing electricity use. Environmental engineers have meanwhile developed devices that help people monitor and regulate their electricity use.

The future of electronics may reside in entirely new systems that have never before been tried. Engineers might soon find ways to mimic the power generation that takes place in nature and thereby design small power cells that do not use electricity at all. A second future opportunity comes from the new science of nanotechnology. Nanoscale materials can be expected to emerge from laboratories for the purpose of ultraefficient energy transmission, reducing energy waste from the power plant to the electrical outlet. Nanoscale materials may eventually be the answer to the design of mini cells for generating power or for storing it.

The electronics industry needs these advances in order to reduce energy demand in a society that is increasingly looking for ways to save. The future of energy-efficient electronics therefore will combine better energy use decisions by people with a new generation of devices, appliances, and equipment.

ECOLOGICAL LANDSCAPE DESIGN

Land degradation represents one of the biggest threats to the environment. The destruction of natural habitat by uprooting forests and grasslands disrupt carbon recycling and storage. New housing developments break up wildlife habitats, cloud streams with excess silt, and cause soil erosion. Even the best-planned developments disrupt nature's landscape. Single houses built on undisturbed land produce similar harms to ecosystems, including the ecosystems that are almost invisible to people, such as soil ecosystems, the species living in foliage, and the microscopic life in nearby streams. Few people could miss the irony that comes from building an energy-efficient solar-powered home on a hillside that has been denuded of trees, bushes, and earth. In order to mesh the innovations of energy-efficient homes with nature, environmental engineers have begun to pay special attention to a new structure's surroundings. For this reason, environmental engineering incorporates a component that considers land topography, climate, and natural vegetation. Ecological landscape design combines the task of coordinating a new building with its natural surroundings and remodeling the landscape to fit the new building in a natural way.

Today's ecological landscape design has three main objectives: (1) minimize the disturbance done to the land during and after construction; (2) minimize all health risks to people and the environment due to the new construction; and (3) reduce construction wastes. At the end of a construction project that has worked within these guidelines, the new structure should function in harmony with the environment. As the theme of this book stresses, environmental engineering seeks to work in cooperation with nature rather than to find ways to subdue nature.

This chapter opens with a description of traditional landscape design that was practiced for years without much regard to habitats or ecosystems. The chapter then explores the ways this older style of landscaping has evolved into ecological landscaping. It also covers new types of architecture that blend structure with nature and discusses the choices and uses of vegetation, soil, watering, and rainwater to enhance a landscape while conserving resources. Finally, special sections examine three related topics: the world's most famous architect, Frank Lloyd Wright; building gardens to preserve biodiversity; and America's scenic highways.

TRADITIONAL LANDSCAPE DESIGN

Landscape design begins with a survey of a specific parcel of land, followed by mapping of all the physical features the surveyors have identified. These features include hills, gullies, trees, streams, ponds, and any preexisting structures, roads, or driveways. A landscaper then contours the land to accommodate any new building about to be constructed on the site.

Traditional landscaping for decades began by cutting down trees and bulldozing the site to make access easier for construction workers. Often landscapers used dynamite to break up large rock formations. They then hauled away surface boulders, filled ponds, and drained wetlands. Sometimes they lopped off the tops of trees to improve a view. These changes to the land were once thought to be enhance-

A trimmed and fertilized lawn became the hallmark of landscaping in the early 20th century. Green lawns appeal to many people, but, unless the landscape is carefully managed, a lawn breaks up natural vegetation, destroys habitat, wastes water, and often includes chemical fertilizers that run off the lawn and contaminate surface waters. Many lawn-care companies now use sustainable lawn management by applying only organic fertilizers and selecting grasses adapted to the environment, so they need less watering. *(Taylor Made Lawncare Services)*

ments to a property, but environmental engineers now understand that the greater the change made to a landscape, the greater the disruption to natural ecosystems. Ecological landscapers have now changed the way in which landscaping is done for three purposes: (1) preserving natural habitat, plants, and trees; (2) leaving as much natural contour in the land as possible; and (3) using the natural landscape to enhance the energy use of the new building.

Today's ecological landscaping still has a need for heavy earthmoving equipment, but ecological landscapers make greater efforts to preserve natural hills and depressions, streams and wetlands, and trees. All landscaping has evolved so that today it incorporates the following aspects:

1. water, topsoil, and energy conservation
2. reduction of storm water runoff
3. horticulture of native plants
4. pest management that reduces the use of chemicals by using natural pest antagonists
5. nutrient replenishment of soils by methods other than chemical fertilizers
6. new plantings and gardens to restore habitat
7. waste reduction and recycling of construction wastes

Landscapers who include these features in their work are called eco-landscapers. Unlike landscaping of just a decade ago, eco-landscaping makes every effort to include the natural geography and topography of the site to be altered. Eco-landscapers follow nature's lead in order to preserve ecosystems and conserve energy.

Even landscapes that contain traditional features such as lawns, gardens with nonnative plants, and nonnative fruiting and flowering trees can be transformed into a more compatible fit with the surrounding ecosystem. Some homeowners have done this by replanting lawns with natural vegetation and replacing nonnative growth, such as a rose garden with a garden of native plants and shrubs. A landscape that in the past received major changes to its contour cannot be rebuilt to its original shape, but owners can take other steps to return the land to a natural state.

LANDSCAPING WITH NATURE

When landscaping with nature, eco-landscapers prepare a parcel of land so that a new building will behave as architects and the homeowners intend, but also reduce its effects on air, land, and water. To preserve air quality, the landscaper designs planting arrangements that prevent dust. Dust-preventing tactics consist of groundcover plants, such as low-growing evergreen shrubs or herbs, ground vines, or pebbles. The landscaper also considers ways to minimize noise from nearby roads or neighbors and reduce disturbance from city lights at night. To preserve the quality of the land, the design should retain much of the natural vegetation, restore ground cover to prevent soil erosion, and avoid landscaping that could lead to mudslides, flooding, or trees downed by severe storms. Water quality preservation can be accomplished by preventing erosion that pollutes streams with silt, retaining the natural course of streams, leaving wetlands undisturbed, and creating a natural buffer zone around flowing waters, lakes, or shorelines.

Biodiversity gardens filled with a variety of native vegetation at varying ages and heights made good artificial habitats for birds, small mammals, reptiles, and amphibians. This type of garden requires clean freshwater in a pond or fountain. Native vegetation gardens need minimal care, and they will not lead to rodent infestation as some people fear because they draw a variety of predators: hawks, owls, and foxes. *(Oregon State University)*

The table on page 136 highlights the main ways that environmental engineers who specialize in landscaping create space that works well for a new home and for the local environment.

Habitats, Inc., is an eco-landscaping company in Eugene, Oregon, that has promoted all of the methods listed in the table on page 136 for building landscapes that blend with nature. The

THE MAIN FEATURES OF LANDSCAPING WITH NATURE		
LANDSCAPING FEATURE	DESCRIPTION	ITS IMPACT ON THE ENVIRONMENT
artificial wetland	provides habitat; cleans runoff; reduces flooding; provides wastewater treatment	contributes to local biodiversity; reduces water pollution
biodiversity garden	native plants and shrubs, including flowering varieties and a water source	provides feeding, shelter, and habitat for native birds, amphibians, reptiles, and insects
buffer zones	undeveloped and undisturbed areas surrounding lakes, ponds, streams, or shorelines	protects existing habitats and ecosystems; reduces erosion; reduces water pollution
drip irrigation	irrigation system that directs a water supply to a specific site, tree, or plant	conserves water; reduces erosion
eco-design	structures built to blend with nature and make nature visible	increases awareness of nature; lessens the impact of a structure on native wildlife behavior
footprint	minimize the total space consumed by a structure and its landscaping	minimizes disturbance to habitats and ecosystems
greenways and open space	undeveloped open space and corridors between open spaces	protects habitats, ecosystems, and wildlife migration routes
native vegetation	plantings and lawns made of varieties native to the local region	encourages plant life suited to local climate, rainfall, soils, and wildlife

Landscaping Feature	Description	Its Impact on the Environment
natural contour	reducing any changes to hills, slopes, rock formations, or streams	reduces overall environment impact; promotes native vegetation and wildlife populations
permeable pavements	porous materials, gravel, or stepping-stone walkways and driveways	reduces runoff; conserves rainwater; nourishes the soil
shading	retaining some trees to shade part of the building and the landscape	blocks excessive sunlight exposure and provides cooling; reduces air-conditioning needs; reduces evaporation
water catchments	roof, garden, or yard receptacles for catching and storing rainwater	conserves water for use in irrigation, supplying water to wildlife, and some home uses
windbreaks	retaining tree stands that provide barriers to wind	reduces heating needs; reduces windblown soil erosion; provides habitat

company expresses the following viewpoint: "Landscapes are part of a living ecosystem, exposed to the elements and forces of nature, and are therefore in a constant state of dynamic equilibrium in the cycles of growth and decay. Creating and maintaining healthy ecosystems reduces the need for intensive routine maintenance." Eco-landscaping therefore has advantages beyond the ones mentioned here. That is, eco-landscaping reduces a property owner's costs and labor because it reduces the need for applying chemicals, pesticides, fertilizers, intense watering, and—because the landscape blends with the natural contours of the land—the need to build retaining walls, drainage ditches, or other structures that try to hold back the forces of nature. The sidebar on page 138 "Frank Lloyd Wright" relates how one architect introduced the concept of designing with nature.

(continues on page 140)

FRANK LLOYD WRIGHT

Study nature, love nature, stay close to nature. It will never fail you.

—Frank Lloyd Wright

Frank Lloyd Wright (1867–1959) was born in Wisconsin but spent most of his early years and professional life in Chicago, Illinois. Wright had dreamed of becoming an architect since his youth, and, after studying civil engineering, he moved to Chicago in 1887 to work for architect Joseph Lyman Silsbee, where he drafted his first building, Unity Chapel. A year later Wright joined Chicago's Adler and Sullivan Architects. Under his mentor, Louis Sullivan, who advocated the philosophy of blending form with function in architecture, Wright expanded his viewpoint to develop the idea of "Form and Function Are One." With Sullivan as a main influence on his budding career, Wright continued designing buildings that had a style never before seen in American design.

Frank Lloyd Wright's buildings in the United States contain several core features that have made them unique and collectors' items in the world of architecture. The most common hallmarks of Wright's designs are: emphasis on the horizontal plane; no basements or attics; no paint; low-pitched rooflines with deep overhangs; uninterrupted walls of windows; liberal use of skylights; large, stone fireplaces in the building's heart; and emphasis on the use of natural materials, such as woods and stone.

As Wright's career and designs progressed, he increasingly integrated nature into his buildings. Sloping roofs matched the slope of the hillsides; large doors and windows lessened the distinction between the indoors and outdoors; and flowing brooks ran under, around, and sometimes through a building. Wright likely became the first person to use *organic* to express a new way of developing and designing an object or even a lifestyle. "Organic buildings are the strength and lightness of the spiders' spinning," he remarked, "buildings qualified by light, bred by native character to everyone and married to the ground." Wright probably never used *sustainable* in connection with his architecture, yet Wright's designs provide excellent guides for present-day sustainable or green houses. Wright followed his own path to sustainability decades before it became fashionable and even necessary. By favoring natural materials, natural cooling and ventilation, and using landscape to create part of a building's structure and support, Wright certainly contributed to today's environmental engineering.

By the end of Frank Lloyd Wright's life, he had developed 1,141 home or building designs and completed 532 of them. His designs and the philosophy behind them influenced architecture in North America, Europe, and Asia. Ten Wright-designed buildings were nominated in 2008 by the Frank Lloyd Wright Building Conservancy to be included on the World Heritage List of the most significant cultural or natural treasures. The United Nations Educational, Scientific and Cultural

Fallingwater in Mill Run, Pennsylvania, was designed by Frank Lloyd Wright. Built between 1936 and 1939 in the Bear Run watershed in western Pennsylvania, the house contained features rarely seen in architecture at the time. Portions of the house appear to float free over a natural waterfall—the house is built directly on top of a stream—and the entire house is in harmony with the woods around it.

Organization (UNESCO) oversees the competition; the nominations and selections will take place between 2009 and 2019. The 10 buildings that have been nominated because they best demonstrate Frank Lloyd Wright's scope, quality, and unique design are the following:

1. Unity Temple, Oak Park, Illinois (completed 1904)
2. Frederick C. Robie House, Chicago, Illinois (1906)

(continues)

(continued)

3. Hollyhock House, Los Angeles, California (1921)

4. Taliesin, Spring Green, Wisconsin (1925)

5. Fallingwater, Mill Run, Pennsylvania (1939)

6. S. C. Johnson and Son, Inc., Administration Building and Research Tower, Racine, Wisconsin (1936, 1944)

7. Taliesin West, Scottsdale, Arizona (1937)

8. The Price Tower, Bartlesville, Oklahoma (1952)

9. Solomon R. Guggenheim Museum, New York City, New York (1956)

10. Marin County Civic Center, San Rafael, California (1957)

The sites listed here and hundreds of other Wright buildings continue to serve as study models in environmental and civil engineering, landscaping, and ecological design.

(continued from page 137)

City residents can also participate in ecological landscaping with a bit of inventiveness. Some city townhouses and apartments include small backyards. Owners can plant these backyards with native shrubs and plants rather than cover the ground with concrete. One native tree fit into even a small yard helps to draw native birds, insects, and other life that bring nature into the city. Plant life additionally removes carbon dioxide (CO_2) from the atmosphere, a process that aids the air quality of traffic-clogged cities. Some city dwellers with a serious commitment to the environment build *permaculture* farms inside their properties. A permaculture farm uses sustainable practices such as composting, rainwater collection, and cultivation of vegetables and fruits. Some city permaculture farms may even raise animals such as chickens if local ordinances allow it.

Cities would benefit from ecological landscaping because the trees and vegetation that are part of an ecological landscape help clean the air and regulate temperatures. But today's metropolitan areas have lost most of their connection to nature so ecological landscaping may not be a practical undertaking. To return even a small parcel of land to the environment requires a landscaper to remove pavement, return streams to their natural shape and condition, and plant native vegetation. The *brownfields* program administered by the U.S. Environmental Protection Agency (EPA) and similar programs in other countries consist of voluntary agreements in which landowners return polluted property to safe use. Sometimes the owner works with the local community to return the cleaned-up brownfield to its natural condition. Once urban brownfields have been restored with natural vegetation and landscaping, they are called green spaces.

ECOLOGICAL ARCHITECTURE

Ecological landscape design must by definition blend a new structure's design with the land. Architect Frank Lloyd Wright in fact stressed that structures are not built on the land but in the land to emphasize the connection between human-made structures and environment. Ecological architecture therefore goes hand in hand with ecological landscaping.

Ecological architecture has the following objectives for the purpose of building a connection between a new structure and nature:

- Buildings conform to a site's natural ecology.
- Buildings maximize the use of natural materials.
- Materials and waste should be managed in a way that returns them to the Earth.
- Materials and wastes do not harm the environment.
- The architecture is based on renewable energy.
- Each building possesses a sense of place that is directly tied to the local environment.

Nature is an overriding theme of ecological architecture, so houses built this way have certain features in common that seem to bring nature right into each room. These features are large windows, skylights, large doors opening onto natural stone patios or a garden, flowing natural water near or in the house, and an emphasis on sunlight during the day and moonlight and the Milky Way at night.

In order to attain a feeling of cooperation between humans and nature, ecological architects create with form and function in mind, often by using

Frank Lloyd Wright's designs as models. Ecological architects also make use of sustainable materials, off the grid energy use, and they have recently begun to explore concepts of biomimicry. Ecological architecture creates designs for buildings that will emphasize the use of renewable resources, natural building materials, and nonfossil fuel energy, all for the purpose of reducing the building's ecological footprint. The ecological footprint is the amount of Earth's land area needed to supply the resources to sustain a person or an activity and dispose of the wastes.

Ecological architecture calls upon new technologies to help meet some of its objectives. The main technologies now being used in ecological architecture are alternative lighting methods, heat distribution, water conservation and reuse, and natural waste decomposition methods. Many of the clues as to how to accomplish these activities may already be part of human history. Ancient civilizations completed magnificent feats of architecture without the use of machinery or steel (but they did make liberal use of backbreaking manual labor). Nature offers its own set of examples on how to maximize lighting, distribute heat and cooling, ventilate, use natural water flow, and decompose wastes.

PLANTS AND TREES

Landscaping with native plants and trees relies on two principles. First, native vegetation requires less input from a landscaper or homeowner because the vegetation has already acclimated to the soil composition, water constituents, and climate. Second, diverse plant life provides a more stable environment than landscaping with a limited variety of plants. A diverse mixture of native plants and trees fosters a diverse mix of insects, birds, and predators, or in other words, a balanced ecosystem. The ecosystem's animal life reduces pests, spreads seeds, pollinates, trims bark, thins out foliage, and recycles nutrients. Variety in vegetation therefore helps maintain animal diversity, which in turn promotes more plant diversity.

Despite the advantages of native trees and plants, people often prefer a manicured look to their property because they have become accustomed to trimmed and lush lawns with neat shrubbery. For this reason, eco-landscapers sometimes blend native vegetation with ornamental plants. For instance, large properties of more than two acres can accommodate both a natural section growing native vegetation and a more manicured section that mixes native trees and plants with ornamental varieties. Part of the property retains its wild appearance, and the other part contains trimmed clearings and periodic removal of underbrush.

Homeowners may also choose to take a step toward sustainability by creating an *edible landscape,* meaning a property containing native fruiting and vegetable plants. The Associated Press writers John Seewer and Doug Whiteman reported in 2008, "The idea goes back centuries, to times when people sustained themselves with food they grew on their own and filled every corner of their land with edible plants. But with the mass production of food, the practice gave way to manicured lawns." There are more than 20,000 edible plants in the world comprised of seeds, stalks, petals,

Edible landscaping involves planting of fruiting and vegetable plants in place of lawn. The plants enrich the soil, reduce erosion, and provide a sustainable use of the landscape. By growing food at home, gardeners decrease the fuel used to drive to a market, and, in a small way, home gardening decreases the energy needed for transporting food around the world. *(iVillageGardenWeb)*

pollen, and roots in addition to familiar fruits and vegetables. The table on page 144 lists some of the most common types that can be used in their native environment as part of an edible landscape.

Edible landscapes can also make use of edible flowers that give color and aesthetic appeal to a yard. The following list summarizes the most common edible flowers used in landscaping today: alliums (leeks, chives, etc.), angelica, anise, arugula, borage, chamomile, chicory, chrysanthemum, clover, dandelion, hibiscus, jasmine, lavender, marigold, pansy, peony, rose, and violet. When growing any edible flower, landscapers should not use chemical pesticides, and they should choose native plants that contribute to the local habitat. The following sidebar "Biodiversity Gardens" on page 145 and "Case Study: America's Scenic Byways" on page 154 explore additional choices in landscaping with nature.

EXAMPLES OF TREES AND PLANTS USED IN EDIBLE LANDSCAPES

FRUIT AND NUT TREES

almond, apple, apricot, avocado, banana, cherry, chestnut, citrus, fig, filbert, guava, kiwi, mango, nectarine, peach, pear, pecan, persimmon, pine nut, plum, walnut

VEGETABLES

asparagus, herbs, horseradish, rhubarb, saffron

GROUND FRUITS

berries (blackberry, blueberry, raspberry, strawberry, etc.), kumquat

VINES

currant, grape, squash

GRAINS AND GRASSES

hops, lemongrass

OTHER

bay laurel, coffee, wintergreen

SOIL, WATER, AND LAWNS

Natural ground cover vegetation such as wild grasses, clover, weeds, and small shrubs helps stabilize the land. Ground vegetation attracts wildlife that replenishes the soil with nutrients at death or by preying on other animals, which leaves body parts on the ground. Some wildlife aerates soil by rooting in search of prey or digging burrows. Natural ground vegetation and trees with strong root systems also conserve water by slowing runoff and evaporation. Despite all the benefits of natural vegetation, many people love the look of a yard blanketed with a green and trimmed lawn,

BIODIVERSITY GARDENS

Land in temperate climates supports various gardens such as ornamental flowers, vegetables, herbs, or cacti. A biodiversity garden acts as another specialized type of garden that helps maintain diversity in nature. Biodiversity gardens contain plants selected to provide habitat, shelter, or feeding for reptiles, amphibians, insects, birds, small mammals, and even invertebrates and microbes. In some urban neighborhoods these gardens may present one of the few places where a healthy ecosystem can function.

In addition to providing for native plant and animal life, biodiversity gardens can serve a role in rescuing plant species that are slipping toward extinction. Two hundred years ago farmers grew a large variety of vegetable plants and fruit trees, but over time commercial farms reduced the number of varieties to make packing and shipping easier. Growers prefer plant and fruit varieties that mature quickly, resist bruising during shipping, and have a long shelf life. In the process of targeting these traits, many thousands of other varieties have become increasingly rare. The Public Broadcasting System (PBS) reported in 1998 that to save plants with a long history in North America, called heirloom varieties, "an increasing number of dedicated gardeners are growing these threatened plants and sharing the seeds with others. These 'seed savers' have rescued hundreds of varieties . . ." The naturalist Craig Tufts of the National Wildlife Federation explained in the PBS show, "When people open their eyes, they can be amazed at what they can find in their own backyards." Biodiversity gardens offer an excellent way for a household to observe nature.

Planning and building a biodiversity garden adheres to the following steps:

1. Explore local open space and determine the types of bushes, shrubs, and plants that grow there.
2. Seek a natural location to plant the garden in a place where it will attract wildlife—near ponds, streams, meadows.
3. Plan vertically as well as horizontally by selecting climbing plants such as ivy or creepers or small trees and build a trellis if needed.

(continues)

(continued)

4. Provide water in a ground-level space and also in a raised birdbath.

5. Include plantings that provide hiding places for birds, reptiles, and small mammals.

6. Substitute organic fertilizers for chemical fertilizers and eliminate use of chemical pesticides.

7. Install raised nest boxes for birds and install nearby bat boxes.

8. Minimize trimming and clipping, but water often.

The Royal Horticultural Society headquartered in London, England, suggests that the best way to assess whether a garden contributes to biodiversity is by watching for birds, bees, and butterflies. Birds indicate that the garden offers nectar, pollen, seeds, and insects and may possibly provide adequate nesting space for a new family of hatchlings. Birds also supply a food source for foxes and hawks. Birds and bats additionally act as effective insect pest control. Honeybees and bumblebees indicate that flowering plants and trees are breeding and producing pollen and nectar. The presence of other insects such as hoverflies and ladybugs indicates that pest control is working because these species eat harmful insects such as aphids. Butterflies contribute to the garden ecosystem because their larvae provide a protein source for birds, small mammals, and other insects.

A good rule of thumb in answering the question of how well a biodiversity garden helps nature is to watch or listen for a variety of insect, reptile, amphibian, bird, and mammal life. Daytime and nighttime activity by a wide variety of species assures that the diverse plantings have successfully provided a space for nature to thrive.

called turfgrass. In some parts of the United States, the perfect lawn can be maintained only through extensive watering, fertilization, and chemicals.

Why do people continue to desire homogeneous expanses of lawn that look nothing like the vegetation that grows in nature? The British geog-

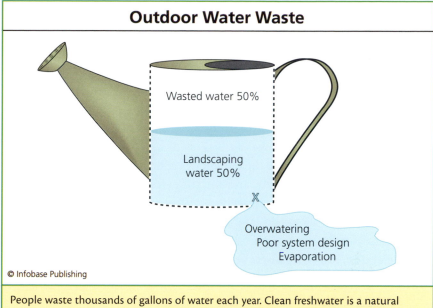

Outdoor Water Waste

Wasted water 50%

Landscaping
water 50%

X

Overwatering
Poor system design
Evaporation

© Infobase Publishing

People waste thousands of gallons of water each year. Clean freshwater is a natural resource that should be conserved like all other natural resources. Building designs and plumbing devices are available to reduce water waste and in some cases to reuse gray water, the wastewater from showers, sinks, and washers. Gray water can then be used for landscape irrigation.

rapher Dennis Cosgrove suggested in his 1988 book *The Iconography of Landscape* that the desire for lawns rather than natural growth is a psychological choice based on how nature was depicted in the arts for centuries. For instance, European landscape paintings from the 17th century show natural land as parks and gardens. The famous gardens of Europe, Britain, and Australia followed this example so that people living in cities could enjoy nature by strolling through trimmed, shaped, and manicured plant life, often sidestepping water sprinklers! Nothing of course could be farther from how nature truly exists. Many people nevertheless have come to think of nature in a parklike sense and feel that natural wetlands, grasslands, and meadows look a bit too messy. Culture therefore has a significant bearing on whether a person chooses a natural landscape over a manicured landscape.

Manicured landscapes involve high-maintenance endeavors that do little to help the environment, mainly because of water waste. The EPA's WaterSense program has highlighted the following factors in water use and its waste in manicured landscape upkeep:

1. People water lawns too often and for too long, oversaturating soil and plants.

2. Many irrigation systems overspray onto paved areas.

3. Weather-based sensors and controllers can reduce water waste on lawns by 20 percent.

4. Soil moisture sensors can determine the amount of water in the ground available to plants.

Turfgrass lawns provide the benefit of reducing soil erosion, but their maintenance, lack of diversity, and thirst for water far outweigh that single advantage. Since eco-landscaping requires strict care in water use and reuse, a new type of natural landscaping has evolved for the purpose of maximum water conservation: *xeriscaping*. In xeriscaping, land contour, plant selection, and irrigation methods all coordinate for the purpose of minimizing water demand. (*Xeri* derives from the Greek word *xero* for "dry"; the word *scape* refers to any view or scene.)

Xeriscaping professionals select native plants with an emphasis on the most drought-tolerant species. In some parts of the United States, such as the dry southwestern region, xeriscaping serves as a crucial water conservation tool. In addition to drought-tolerant plantings, shade trees take part in a xeriscape by preventing excess water loss through evaporation. In some instances, xeriscaping also needs soil amendments so that the natural soil holds water better and does not drain too quickly. Organic matter

The plants selected for a landscape should be native or adapted to the climate. Drought-resistant plants such as these help reduce water waste. (A) red columbine (B) coral honeysuckle (Trees 4NC)

and humus added to fast-draining soils retain moisture, and sand helps slow-draining soils drain faster and prevents rotting. Landscaping therefore offers homeowners a variety of techniques for conserving water.

RAINWATER HARVESTING

Rainwater harvesting consists of the collection of rain in rooftop or ground vessels so that the water can be used in irrigation when needed or shunted to a safe water source for indoor use. Collected rainwater, along with gray water and reclaimed water, represents an alternative water source that helps reduce overall household water bills by 25 percent. Gray water is the excess water that runs into household drains connected to sinks, showers, bathtubs, and washing machine rinse cycles. (Blackwater comes from the same sources, plus toilets, and represents any wastewater expected to contain disease-causing microbes.) Water conservation systems pump gray water to holding tanks for later use in irrigation. Reclaimed water is water that has been treated for uses other than drinking, such as treated wastewater from dishwashers.

Rainwater collection systems use gravity to do the work, making these systems easy to use and inexpensive. Rainwater collects into large cisterns, barrels, or rooftop tanks connected to a downspout that carries it to a holding tank on or under the ground. The few precautions that should be taken in managing rainwater collection systems are the following:

1. a cover to prevent entry of animals or mosquito breeding
2. fencing or other security measures to prevent small children from falling in
3. state-by-state differences in laws regarding the use of rainwater

Alternative water sources additionally play a role in xeriscaping by making use of almost every drop of water in an area to both aid the environment and conserve water use and costs.

MICROCLIMATES

A microclimate refers to a specific small area that has a climate different from the larger area around it. Eco-landscapers who work in regions that have many diverse microclimates must understand these differences in

Xeriscaping

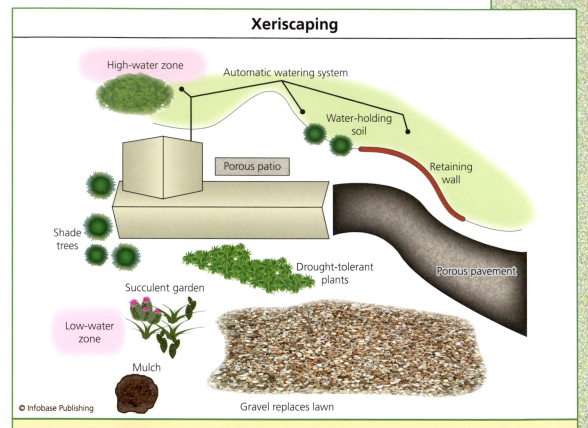

High-water zone

Automatic watering system

Water-holding soil

Porous patio

Retaining wall

Shade trees

Drought-tolerant plants

Succulent garden

Porous pavement

Low-water zone

Mulch

Gravel replaces lawn

© Infobase Publishing

Xeriscaping incorporates the best techniques for reducing water waste; xeriscapes often occur in very hot, dry climates. Irrigation covers any area that needs extra water, but it does not waste water on drought-resistant plants. Porous materials for driveways, walkways, and patios allow rainwater to run into the ground, as does the replacement of lawn with gravel.

order to select the best plants for a particular parcel of land. Some microclimates spread over several acres, but other microclimates are so distinct, they cover no more than a half-acre.

The U.S. Department of Agriculture (USDA) publishes a "Plant Hardiness Zone Map" that shows North America divided into 21 distinct climate zones and helps landscapers and gardeners understand their unique growing conditions. Each different zone on the map contains conditions suitable for certain plants, and landscapers are wise to select plants that do best in these zones. But even within such discrete climate zones, much smaller microclimates often exist. The San Francisco Bay Area in California is a champion of sorts in microclimates: It contains 20 major microclimates and perhaps hundreds of small microclimates across its nine counties. In less than an hour, a person can drive from a hot, dry 100°F (38°C) climate to foggy and windy 50°F (10°C) conditions. During that journey, the driver will likely pass through rain, sunshine, and varying degrees of heat and cold. The *San Francisco Chronicle* writer Harold Gilliam explained in 2001, "The reasons for this extraordinary variety of weather lie deep in the geologic past—in the clash of tectonic plates. The Pacific plate of the Earth's crust, moving eastward over the eons, smashed into the edge of the North American continental plate, prying it up into monumental chains of mountains, including the Sierra Nevada, and, much later, California's coast ranges. It is this rumpled landscape, this hill-and-valley topography, this heterogeneous diversity of landforms that give the Bay Area its multiple climates, microclimates, and submicroclimates." Each eco-landscape therefore consists of a design best suited to its microclimate.

Environmental engineers must understand geography and climate when seeking energy efficiency because these things contribute to microclimates, which in turn influence the efficiency of buildings and landscaping. The following list summarizes the factors that engineers consider when working with a microclimate:

1. path and intensity of sunlight
2. wind and rain directions
3. wind forces
4. barometric pressure patterns
5. morning, midday, and evening temperature and humidity
6. coldest and warmest temperatures by season
7. cold air drainage
8. soil evaporation rates and tree transpiration rates
9. bodies of water that absorb daily heat and moderate temperatures
10. north- or south-facing slopes
11. soil composition, chemistry, and moisture

Microclimates and all the many factors they encompass explain why environmental engineers require training in diverse sciences in order to work within the natural boundaries of nature.

WALKWAYS AND DRIVEWAYS

Rainfall that seeps into the earth receives cleaning as it percolates past soil particles that remove chemicals and microbes. The water then trickles downward into natural aquifers that serve as underground reservoirs for clean drinking water. When communities pave over the ground for roads, driveways, and parking lots, rainwater can do nothing but rush downhill into streams and then to wastewater treatment plants. This event has two drawbacks: Rainwater does not become available for plants and aquifers, and the extra volume of water requires extra energy consumption by wastewater treatment plants. Stress put on treatment plants can be significant in heavy storms, and the incidence of drinking water contamination increases when storms overtax local wastewater treatment facilities. For these environmental and health reasons, permeable pavements offer a useful way to manage and conserve water.

Porous surfaces act as one of the best ways to reuse rainwater and irrigation water. They also reduce runoff to streams. (A) An ecological garden uses stones to create a pathway that allows water to enter the soil. The gardener has planted thyme between these stones. *(Kurt Lawton)* (B) A pebble driveway helps in water recycling much more than a paved driveway. *(Bushnell House)*

A permeable pavement is any material that provides pedestrians or vehicles with a strong surface but also allows water to drain through rather than run off the surface. Landscapers select from five main types of permeable pavements, described in the following table.

Permeable pavements must be selected with the soil conditions in mind. For example, clay soils do not drain well and have a tendency to turn to mud in heavy rains. Areas with a very high water table, meaning the area's natural groundwater storage is close to the surface, present the opposite problem: Permeable materials allow too much water to enter already soaked conditions. Even the best-designed permeable pavements can clog and may need extra maintenance to keep them open.

PERMEABLE PAVEMENTS		
PAVEMENT TYPE	**ADVANTAGES**	**DISADVANTAGES**
porous asphalt	uses less petroleum product than non-permeable asphalt	does not hold up in high traffic/high speed roads
porous concrete	lowers solar heat gain by city centers because of its light color	concrete is a energy-intensive product to make
plastic grids	high strength and provides a way of recycling waste plastic	often require another support layer above or below
block or stone pavement (brick, stone, gravel)	appealing looks	expensive and gravel can erode and clog water drainage systems
stone lattice	lattice structure allows ample drainage and promotes the growth of grass in the openings	may require extra maintenance

CASE STUDY: AMERICA'S SCENIC BYWAYS

Since 1991 the U.S. Department of Transportation (DOT) National Scenic Byways Program has designated specific roads that lead to archaeological, historical, cultural, natural, recreational, or scenic locations. The DOT has so far designated 125 such roads in 44 states. All of the scenic byways encourage travelers to gain greater appreciation of the land and its local trees, plants, and wildlife. Because scenic byways often call attention to nature, they seem to be a perfect place to adopt eco-landscaping. But heavily traveled roads have also caused harm to biodiversity by breaking up habitats, blocking migration routes, and killing wildlife in road accidents. Current and future scenic byways hold an important opportunity to establish ecologically friendly methods to construct and maintain popular roads.

Scenic byways range in rustic quality from the Beartooth Highway that meanders through Montana's mountains and forests to Delaware's Brandywine Valley Scenic Byway that travels past large estates and magnificent botanical gardens. Regardless of the terrain that a scenic byway traverses, the program ensures that all portions of the highway minimize the presence of billboards, telecommunications towers, fast food restaurants, or other clutter. One goal of the scenic byways program involves saving open space, and keeping human activities away from the roads helps preserve open space. In many instances, scenic byways probably provide some people with a look at nature that they would otherwise miss entirely in their urban lifestyle.

LANDSCAPE DESIGN SKILLS

The landscape design profession combines art and science for the purpose of planning and shaping the land that surrounds a structure. Ecological landscape design focuses on meeting a customer's needs while also planting trees and plants, gardens and ponds, and other features that create a relationship between the building and the environment. Landscape designers must understand many of the same principles as environmental engineers, that is, hydrology, geography, topography, and climatology. The landscape designer, however, emphasizes the artistic use of plants, water, and land. The following table summarizes the basic skills that landscape designers call upon when they develop a new but natural look for a parcel of land.

LANDSCAPE DESIGN SKILLS	
THE MAIN ELEMENTS OF ART	
color	understanding of the relationships among primary, secondary, and tertiary colors; complementary colors; tint (lightness of a color)
line	creation of a structure that causes eye movement or flow, usually related to a structure's outline
form	shape and size of an object: large or small circle, oval, square, rectangle, or triangle; columnar, spreading, creeping, or weeping shapes
texture	surface quality of an object: smooth, coarse, prickly, glossy, dull
scale	size of an object in relation to adjacent objects
THE MAIN ELEMENTS OF DESIGN	
unity	consistent style and effective use of objects to convey a main idea
balance	symmetrical or asymmetrical designs
transition	gradual change in color, size, or other characteristics
proportion	relationship of objects' sizes to each other and to the overall design
rhythm	arrangement of objects creates a feeling of motion and guides the eye in a direction
focalization	arrangement of objects to draw the eye toward a central point, the focal point
repetition	repeated use of similar features
simplicity	elimination of unnecessary detail or clutter

Environmental landscaping begins with a plan agreed upon between a house's architect and the property's landscaper. A landscape designer also gives input to this process in order to create a visual meaning to an entire property. For instance, a country cottage would look out of place nestled among large rock formations and towering pines, and the property would be further confused by a modernistic landscape design including sculpted shrubbery and enormous fountains. Landscaping in harmony with nature eliminates most of these pitfalls: Nature always produces the best balance between art and design.

The landscape design process entails the following six steps:

1. Develop a design for a given parcel of land.
2. Conduct a site analysis to assess existing tree and plant life and physical features.
3. Assess the house design, landscaping objectives, and the residents' desires.
4. Locate areas of the property that will be designed.
5. Create design plans for those individual areas.
6. Select the plants to be used and plant them in the designated areas.

In order to carry out the tasks listed here, landscape designers depend on strong skills in horticulture. In this way they help eco-landscapers select plant life that blends with the following factors: local climate and any microclimates; soil conditions; wind and storm incidence; and potential freezing or drought. Landscape designers today also design biodiversity gardens and edible gardens.

CONCLUSION

Ecological landscape design provides ways in which a house or other structure can work in concert with the local land, vegetation, and climate. In the simplest sense, this field of expertise blends the needs of people with the needs of nature. Eco-landscapers avoid older landscaping methods in which forests or meadows were cleared and leveled for new developments. Eco-landscapers today are more likely to fit a new construction into the existing shape and vegetation of the environment.

This helps maintain ecosystems even while a new building construction takes place.

The process of building a new structure that works with the land requires three related fields of expertise: eco-landscaping, ecological architecture, and landscape design. Very often the same professional performs both the landscape design and its actual landscaping. All of these skills require the willingness to cast aside older, traditional concepts of design and adopt new types of form and function. Not only must architects and landscapers make this transition in thinking, but homeowners also must realize the advantages of a nonconventional-looking home that works with nature to save energy.

Today engineers, architects, and designers pay close attention to the characteristics of a specific piece of land to be developed. They then create innovative ways of designing a structure that does not harm the local ecosystems. To do this, eco-landscaping creates structures that work with the environment, but also draws upon nature's shapes, sizes, materials, and modes of function. Ecological landscape design is not a luxury that only the wealthy can employ for trendy new homes or eye-catching commercial buildings. Ecological landscape design is instead a basic need for future sustainable communities. Like nature itself, these communities will rely on efficient energy processes that use sources other than nonrenewable fuels, and they will produce wastes that degrade in the environment without poisoning the air, water, or land.

SUSTAINABLE WASTEWATER TREATMENT

ydrology, the science of water systems, has always been a central part of environmental engineering. Ancient societies used flowing water as a means of travel, a conveyance for moving timber and other goods downstream, and a waste disposal system. During the growth of civilization, people learned that bodies of water held a significant amount of force. Villages constructed mills next to fast flowing rivers to let the force of the water supply power for the milling operations. The next generation built enormous dams to control the force of water that powered turbines and produced electricity. Water systems will become part of new energy-generating systems in sustainable communities in the near future, mainly because water avoids the use of nonrenewable and pollution-causing fossil fuels.

Wastewater treatment follows a standard process that turns raw sewage into disinfected water safe for return to the environment. Almost all modern wastewater treatment plants conduct the following steps to treat wastewater: (1) wastewater passes through screens that remove large solids; (2) wastewater enters a grit chamber where heavy wastes settle out of the water by gravity; (3) water enters a larger settling tank where light, small particles slowly settle out of the water; (4) the water enters an aeration tank containing aerobic (oxygen-requiring) bacteria that digest organic substances, aided by a constant bubbling of air through the contents; (5) the water enters another settling step and then passes through a filtration tank, which removes very fine particles; (6) disinfectant kills the bacteria in the water; and (7) the treated water discharges into the environment.

Pipes carry much of the heavy organic sludge that collects in the settling tanks to another tank called an anaerobic digester. This digester contains anaerobic (requiring the absence of oxygen) bacteria that slowly degrade the heavy sludge and produce methane and small amounts of other gases. The wastewater treatment industry calls the methane biogas, and both biogas and solid biomass serve as energy sources produced by wastewater treatment.

This chapter discusses the transformation of wastewater treatment systems into dual purpose waste treatment–energy production facilities. It covers modern waste-to-energy plants and explains the relationship between water and energy. It also describes gray water reuse and provides insight into how developing countries are finding ways to take advantage of wastewater treatment. This concluding chapter explains one of the most important tasks in environmental engineering today: sustainable wastewater management.

THE ENERGY-WATER CONNECTION

Environmental engineers apply the first and second laws of thermodynamics in every project they plan and develop. The first law of thermodynamics states that energy can neither be created nor destroyed. The form of energy, however, can change. For example, the motion energy inside a turbine changes to electricity. The second law of thermodynamics states that energy flows from a region of high concentration to a region of lower concentration, much as water flows downhill. By the second law of thermodynamics, water at the top of a waterfall contains a high concentration of energy. As it pours over the waterfall and fills a basin, the energy concentration changes. The water that settles into the basin at the bottom of the waterfall holds a lower concentration of energy. This principle has been used for centuries to power daily operations.

Flowing water illustrates another component of energy—potential energy versus kinetic energy. Water at the top of the waterfall before flowing over the edge contains potential energy, that is, a form of stored energy. As the water pours downward it loses potential energy and gains kinetic energy, the energy of motion. All of the different types of energy

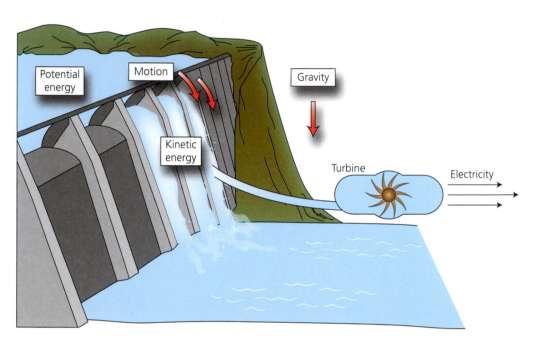

Water and Energy

Operations that generate energy from flowing water—hydroelectric dams, wastewater treatment plants, mills—rely on capturing water's kinetic energy. Water at the top of the waterfall in this diagram holds high potential energy, but after it reaches the bottom of the waterfall it contains low potential energy. Power plants take advantage of this transition.

in the solar system can be classified as either potential or kinetic, as the following table shows.

Energy can change from one type to another, as mentioned, illustrated by the following example:

1. A racehorse converts the chemical energy in hay to motion energy.

2. A person converts electrical energy to thermal energy by turning on an oven.

3. Plants convert radiant energy from the Sun into chemical energy through photosynthesis.

TYPES OF ENERGY	
KINETIC ENERGY	**POTENTIAL ENERGY**
electrical—movement of electrical charges carried by electrons	chemical—energy stored in the bonds in molecules
radiant—electromagnetic waves, such as light or solar energy	mechanical—energy stored in an object under the application of force (example: a bowstring pulled before releasing an arrow)
thermal—heat created by fast movement of atoms and molecules within matter	nuclear—energy held in an atom's nucleus
motion—movement of objects from one place to another	gravitational—energy held by an object because of its placement (example: a boulder at the top of a hill before rolling)
sound—movement of energy in vibrations or longitudinal waves called compression/rarefaction	

Water's use as an energy source throughout history has usually been as motion energy. Wastewater treatment plants can take advantage of the natural flow of water to power operations, but wastewater offers an additional energy source in the form of biomass. Biomass is organic matter in wastewater that comes from plant and animal wastes. In the wastewater treatment industry, biomass represents a source of chemical energy that can be converted to other forms such as thermal energy.

WASTEWATER IN DEVELOPING COUNTRIES

Numerous developing parts of the world contend with a high incidence of infectious disease because of poor drinking water quality and poorly functioning wastewater systems. Leaks and contamination of either system come from faulty or missing infrastructure, meaning distribution pipes, collection pipes, treatment facilities, and disinfection methods. Because of serious health threats to humans, many developing countries that already struggle with poor infrastructure make wastewater treatment their priority. The World Health Organization (WHO) has warned that the problem will get worse in developing countries. Because these places have clean water shortages, wastewaters substitute for treated water for irrigation, fishing, and cleaning clothes.

Sustainable wastewater treatment refers to the production of energy from the wastewater treatment process. This method of energy production offers an opportunity for developing countries for two reasons: the efficiency in converting biomass's chemical energy to other forms of energy, and low cost compared with other energy technologies. Sustainable wastewater treatment therefore relieves two problems that plague many developing parts of the world: energy and waste management.

Nonindustrialized countries have a great need for inexpensive, easy-to-build waste treatment technologies. Some of these countries have adopted inventive and sustainable solutions to waste treatment. The Internet reporter Andrea Millar provided an example in a 2008 article: "A Kenyan jail has confronted the issue of its waste production by creating a sustainable wastewater treatment facility staffed by inmates and designed by both local and international organizations. Rather than employ high-tech engineering, the bulk of the [treatment is] handled by natural processes in the facility's man-made wetlands. Acting as a vast filtration system, the sustainable processes of anaerobic microbes convert human waste into materials for biogas and water for use on the jail's vegetable fields." Millar's example illustrates a situation in which wastewater handling is done in a manner that requires no energy input at all other than a small amount of manual labor.

The next step in sustainable wastewater use involves methods for producing energy. Treatment plant anaerobic digesters offer the most efficient way to make energy-containing methane gas, but some villages may not have the funds to build a new plant from the ground up. A waste treatment pond offers an inexpensive option because it contains conditions similar to those found inside treatment plant anaerobic digesters. Treatment ponds contain anaerobic activity in the deepest parts of the pond where aerobic bacteria have depleted the oxygen. Anaerobic bacteria live naturally in these types of places so setting up a wastewater treatment pond carries little expense. Methane gas drifts up from the bottom of the pond and can be captured and piped to an energy-generating plant so that the pond produces energy rather than consumes it.

Some ponds tend to grow heavy mats of algae on the water surface. Rather than spend efforts to get rid of the algae, workers remove the algae layer and add it to a small anaerobic digester to produce heat. In fact, treatment ponds provide the same advantages of mechanical anaerobic digesters: They help in waste management and treatment and recycle nutrients. One disadvantage of ponds compared with closed anaerobic digesters comes from odors emitted by the bacteria in their normal digestion of organic matter. The odors do not cause a health problem, but smell unpleasant.

R. Otterpohl wrote an online article in 2007 for the Swiss Federal Institute of Aquatic Science and Technology describing new processes for wastewater treatment that seem suited for developing countries: "The conventional wastewater management concept, consisting of a waterborne wastewater collection system leading to a central treatment plant, has been successfully applied over many decades in densely populated areas of industrialized countries . . . However, the appropriateness of this model in the context of developing world cities must be questioned, given the urgent need for affordable, sustainable infrastructure. During the last decade, various researchers and institutions, including the World Bank, have started to consider the decentralized wastewater management approach as an alternative to conventional centralized systems, but these approaches have struggled to gain acceptance." Part of the resistance comes from the worry that treatment plants intended for the use of only one or two towns might not be as efficient to run as big centralized wastewater treatment plants. But small, decentralized treatment plants also offer advantages to communities, as follows:

1. smaller treatment systems tailored to a community's specific needs
2. experimentation with different wastewater treatment methods
3. reduces accidents that may occur in long-distance wastewater shipment
4. increases opportunities for treated water reuse
5. makes incremental changes and improvements manageable

Today, the Engineers Without Borders-International (EWBI) organization helps communities in developing regions adopt new technologies to improve their health, income, and lives. The EWBI stresses sustainable methods in energy use, water management, and native natural resource conservation. EWBI engineers and other volunteers help disadvantaged communities build needed structures and teach residents basic engineering skills for future projects. The founder of the Engineers Without Borders-United States, Bernard Amadei, revealed his view of the organization's purpose in a 2007 interview with the correspondent Spencer Michels. Amadei described San Pablo, Belize, as a village "where I noticed a lot of little girls, young girls, who were carrying water—that was their job—from the river to the village, back and forth, back and forth. And as a result, they could not go to school. It broke my heart. And I decided I was going to do something about it." EWBI develops projects with a community, not for it, so that the community solves its own environmental engineering needs. EWBI helps train residents how to build a structure, how to fix it, and how to keep it running.

Developing countries therefore have options for the type of wastewater treatment they choose. Because of the health needs for safe wastewater management and the need to produce inexpensive energy, sustainable wastewater systems are ideal for developing parts of the world. The following "Case Study: Kufunda Learning Village, Zimbabwe" illustrates how

CASE STUDY: KUFUNDA LEARNING VILLAGE, ZIMBABWE

Kufunda Learning Village in Zimbabwe, located near the capital of Harare, was created in 2005 by Marianne Knuth for the purpose of creating strong, healthy communities that are responsible for their own future. In addition to cultural growth, Kufunda has embarked on an exploration of technologies to meet its most immediate needs. Kufunda residents focus on building self-sufficiency in two main areas: agriculture and community infrastructure. Knuth explained, "We are a small group of people who have decided to create a learning village aimed at the creation of locally rooted solutions to community self-reliance challenges. A starting assumption for my work here is that people already know how to work in creative and self-sufficient ways, and that the challenge is to help them access that knowledge—and the self-confidence to act on it—generating concrete and often surprising results in the process." In time, Kufunda's residents hope they can transfer their knowledge to neighboring villages that wish to create similar self-sufficiency.

In the face of periodic political unrest, villages like Kufunda have a greater need for self-reliance, perhaps, than any other communities. Residents have begun to investigate the develop-

single communities can build their own sustainable lifestyles by adapting technologies and applying simple effort.

ANAEROBIC DIGESTERS

An anaerobic digester is any equipment or site that holds waste materials in an oxygen-free environment so that anaerobic bacteria can degrade the wastes. On a wastewater treatment plant's property, the digester is a large tank that holds several thousand gallons of liquid and semiliquid material. Wastewater digestion also takes place in a similar way in lagoons or ponds even though these places are exposed to the air. Though the upper layers of the water contain some dissolved oxygen, the deeper layers become less aerobic and more anaerobic. If the contents of an anaerobic digester or pond receive little mixing, oxygen does not penetrate the depths and so anaerobic bacteria work at their best.

Anaerobic decomposition occurs naturally in swamps, bogs, stagnant ponds, deep bodies of water, and waterlogged soils. The bacteria in these places degrade organic matter to the simplest of compounds, which allows

ment of ecological projects such as renewable energy, organic farming, ecological building, and composting toilets. The Kufunda Learning Village Web site has reported, "Ecological sanitation is a system that makes use of human excreta and turns it into something useful, which can be used to grow plants or trees. By now all the toilets at the village are of this type. Our simplest compost toilet can now be built for roughly 10 U.S. dollars—an important factor when working with financially poor rural communities." Not only are Kufunda's plans cost-effective, but they make every use of natural clean-running systems.

The Kufunda Learning Village has already begun to run its main electronic systems on solar power—the satellite Internet was the village's first system to run on solar energy. Kufunda's residents have an advantage over richer, more industrialized communities: They own no preexisting energy or waste management systems to tear down. The Kufunda Learning Village continues to find the best solutions to its specific environmental issues and, in the process, provides a good example of building efficient and simple technologies that foster sustainability.

the nutrients to be recycled. For example, amino acids degrade to carbon dioxide (CO_2), methane (CH_4), hydrogen (H_2), and small amounts of nitrogen and sulfur compounds.

Digestion of organic matter—in nature as well as in a digester—takes place in three stages. First, aerobic bacteria degrade complex compounds such as starches, proteins, and fibers into smaller carbohydrates or peptides. A second group of bacteria use these compounds as food and produce organic acids as end products of their enzyme reactions. The organic acids all have in common a carboxylic group as part of their structure (COOH). This section of the molecule contains a carbon that is linked both to an oxygen molecule and to an oxygen-hydrogen complex, called a hydroxyl group. Examples of the organic acids produced in this step are acetic acid (two total carbons), propionic acid (three carbons), and butyric acid (four carbons). As a third and final step, anaerobic bacteria use the acids for energy and produce methane, carbon dioxide, and a small amount of other gases (hydrogen, carbon monoxide, hydrogen sulfide, and nitrogen).

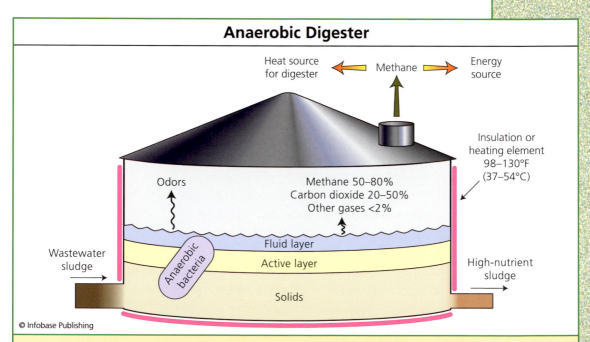

Anaerobic Digester

Heat source for digester ← Methane → Energy source

Insulation or heating element 98–130°F (37–54°C)

Odors

Methane 50–80%
Carbon dioxide 20–50%
Other gases <2%

Fluid layer

Wastewater sludge

Anaerobic bacteria

Active layer

High-nutrient sludge

Solids

© Infobase Publishing

Anaerobic digesters at wastewater treatment plants make fuel in the form of methane gas, which can be burned to release its energy. Many sustainable wastewater treatment plants use the methane to provide heat for the digester so that the entire process can continue. This is an example of a sustainable loop.

The methane and the other biogases produced by anaerobic digestion serve as an energy source that can be used to run the treatment plant or do other work. Burning one cubic foot of biogas yields 10 Btu of heat energy. The amount of methane in the biogas mixture affects the energy production because methane is the main energy source. Each percentage of methane results in 10 Btu, so that a biogas containing 65 percent methane produces 650 Btu per cubic foot. Very active anaerobic digesters can produce enough methane to run a treatment plant's heating, refrigeration, and electricity. Anaerobic digesters therefore play a role as the main powerhouse when designing a sustainable wastewater treatment plant.

Methane cannot solve every environmental problem. For one thing, though methane can be used for making energy, it is also a predominant greenhouse gas that causes global warming. Sustainable wastewater treatment plants help reduce this problem by capturing all the methane emitted by a digester and using it, but methane come from additional sources in the world. Human activities that lead to methane production are waste treatment, biomass burning facilities, energy plants, and landfills. Methane also comes from natural sources: anaerobic soil and water bacteria, some vegetation, fossil fuel deposits such as coal mines and natural gas fields, and certain animals. Scientists have begun to realize in the past decade or so that the animal sources of methane contribute a significant amount of greenhouse gas to the atmosphere. The sidebar on page 168 "Methane—Cow Power" discusses this type of natural methane production and a few opportunities that may come from it.

GRAY WATER REUSE

Gray water is water that has been used for showering, faucet uses, and clothes washing, and it usually disappears down drains as a wasted resource. Sustainable wastewater treatment involves the recapture of gray water for additional uses, a process better known as water reclamation. These recovery systems do not collect kitchen sink water or toilet water because these can be health hazards due to the presence of disease-causing microbes.

Sustainable houses collect gray water in pipes that run separate from toilets and the kitchen sink. After collection, building designers have options regarding how and where the gray water will be best used. The three main options for gray water use are: (1) as irrigation water; (2) for flushing toilets; or (3) in the home's fire suppression system.

METHANE—COW POWER

Cattle, meaning dairy cows and beef animals, belong to a diverse group of animals called ruminant animals, or simply ruminants. Ruminants possess a four-part digestive system composed of a stomach similar to a human stomach and three additional chambers that carry out a different digestion than found in the stomach. Ruminants exist on a diet of fibrous plants that require a good deal of chewing to digest. These animals do this by swallowing partially chewed food, regurgitating it, chewing it again to further break up the pieces (called chewing the cud), and then re-swallowing the material. The smaller, chewed pieces move into the other compartments that contain enormous numbers of anaerobic bacteria and protozoa, especially the biggest compartment called the rumen. Rumen microbes degrade fibers into organic acids that the animal relies on for energy. As the microbes work they release methane, which the animal must constantly eliminate by belching or passing along to the large intestines. The methane drifts upward into the atmosphere and mixes with other greenhouse gases.

CO_2 receives most of the blame for the crisis of greenhouse gases and global warming, but, in fact, other gases cause much more harm to the environment. Methane contributes about 20 times more to global warming than CO_2. The EarthSave environmental organization has calculated that animal agriculture produces 100 million tons (91 million metric tons) of methane a year. Cattle produce about 20 percent of the methane emissions in the United States, and, unlike new methane-reducing technologies for factories and wastewater treatment plants, these cattle must keep producing methane as long as they live.

Environmental scientists have considered two divergent solutions to cattle methane: capturing the methane for energy or altering normal ruminant digestion to produce less methane. Capturing the methane from belching cows presents an understandably difficult task, so engineers have turned their attention to the large amounts of methane that come from cattle manure. Pacific Gas and Electric Company has embarked on a program to capture this "cow power." Roy Kuga, its vice president of energy supply said, "With nearly 2 million dairy cows in California, there is great potential for the state's agriculture and power sectors to work together to address the

Two new technologies may soon join the three main uses for gray water. The first involves the routing of gray water into wetlands constructed near the building to naturally degrade liquid wastes. These *constructed wetlands* work similarly to natural wetlands in which wastes move very slowly through the site so that plants and microorganisms have time to degrade the organic matter. Lori Ryker explained in her 2005 book *Off the Grid*,

challenges of climate change." Cattle have become a hot topic among environmental scientists due to their enormous production levels of a renewable energy source.

Cow power farms work by conveying the manure to covered tanks or lagoons. Gas collecting devices receive the methane as it rises out of the waste load and, with a small amount of processing, pipe the natural gas to an energy utility company's distribution system. On the other side of the country, Central Vermont Public Service, an electric utility company, uses manure methane to fuel an energy generator and converts the output to electricity for its customers. A Vermont environmental planner Jason Bregman has predicted, "The next generation of renewable energy systems will seek out organic matter in municipal, commercial and agricultural waste streams as a relatively easy source of fuel to obtain and process energy." Other states and many other countries have begun their own methane waste-to-energy programs.

Reducing the methane a cow produces requires adjustments to the cow's diet. Grasses that cattle graze contain a stringy fiber called lignin that can only be digested with cud chewing and the action of the rumen's anaerobes. Laboratories have begun to develop new grasses that contain less lignin so that grass-fed beef and dairy cows emit less methane. The Carnegie Mellon University engineer Christopher Weber explained to *Discovery* magazine in 2008, "Genetically modified grass could be an appealing solution. It could be more acceptable to the carnivores among us than meat grown in test tubes or giving up meat altogether." The promise of adjusting cattle's diets offers some promise, but it is still a long way from helping solve climate change. A more daunting obstacle comes from the fact that cattle are not the world's only methane producers. The following ruminant animals all produce methane: alpacas, antelopes, bison, camels, deer, giraffes, goats, llamas, oxen, pronghorns, sheep, water buffalos, wildebeests, and yaks.

Methane use and methane reduction are two rapidly emerging technologies. If environmental engineers can design successful means of capturing methane for energy use, they will certainly help in advancing sustainable energy sources. Overall, methane will continue to be a problem in global warming.

"The goal of a man-made wetland is to replicate a natural wetland's ability to clean and filter water. A constructed wetland system's selected plants filter gray water in a specific order, and then return the water to the earth, eliminating the need for a fully developed septic system." The second technology uses very efficient filtration systems to make the water suitable for drinking, known as *potable water*. The following filtration technologies

A sustainable home manages three different water sources: (1) rainwater, which is usually captured in a cistern such as the 60-gallon (227-l) size shown here; (2) gray water, which is wastewater from showers, sinks, and washers that is not expected to contain disease-causing microbes; and (3) blackwater, which is wastewater from toilets, sinks, or any other source expected to be contaminated with disease-causing microbes. *(Natural Rainwater.com)*

may soon produce drinking water for direct reuse as potable water: ultrafiltration in which filters contain extremely small pores to capture all contaminants; nano-technology in which nanoscale materials act to clean or even sterilize water; and reverse osmosis in which water is forced through a fine membrane to clean out all impurities. Another useful type of filtration for water purification is described in the sidebar "Carbon Adsorption" on page 173.

Most houses that recover gray water today send the water to a holding tank that degrades waste in a manner similar to a septic tank. The cleaned water moves out of the tank by passive energy; it is pushed along by the force of new gray water entering the system. The cleaned water flows through a short series of baffles that prevent most of the dirty gray water from mixing with the cleaner water, and then the cleaned water goes through the outflow pipe. After completing this cleanup process, the two main current uses for gray water are garden irrigation or flushing toilets. Gray water reclamation does not need to be confined to houses; these closed-loop systems can work in large office buildings, schools, and manufacturing plants.

ECOLOGICAL WASTEWATER TREATMENT

Ecological wastewater treatment encompasses three control methods for keeping wastewater from harming the environment: odor control, methane control, and discharge cleanliness. Odors from gases like hydrogen sulfide emitted from anaerobic digestion do not harm human health but make the surroundings unpleasant for neighbors of treatment facilities. Anaerobic digesters help solve most of this problem, and methane collection devices over manure tanks also reduce odors. Methane collection as previously discussed in this chapter is a critical part of ecological wastewater treatment because it can have a meaningful effect on greenhouse gas levels. The third factor, discharge of cleaned water, represents the wastewater industry's major responsibility and the U.S. Environmental Protection Agency (EPA) enforces strict laws on the quality that treated wastewater must achieve. The government expects treatment plants to keep the amounts of treated wastewater constituents within certain predetermined limits. The following table describes the constituents in wastewater that the EPA monitors to avoid harm to the environment. These constituents must be controlled within the EPA's acceptable limits whether a treatment

IMPORTANT WASTEWATER CONSTITUENTS THAT WASTEWATER TREATMENT MUST CONTROL		
CONSTITUENT	DESCRIPTION OR SOURCES	COMMON CONTROL METHODS
suspended solids	organic and inorganic insoluble matter	grit screens; sediment settling tanks; clarification tanks; flotation; coarse matter filtration; chemical precipitation
biodegradable organic matter	human and animal waste; food waste; leaves, grass, and other organic matter in runoff	aerobic bacterial digestion; anaerobic bacteria digestion; physical-chemical clarification; fine matter filtration

(continues)

IMPORTANT WASTEWATER CONSTITUENTS THAT WASTEWATER TREATMENT MUST CONTROL
(continued)

CONSTITUENT	DESCRIPTION OR SOURCES	COMMON CONTROL METHODS
nitrogen compounds	human and animal waste; drugs and hormones	chemical oxidation; ion exchange
phosphorus	human and animal waste; drugs and hormones	chemical treatment; biological phosphorus removal
pathogens	disease-causing bacteria, algae, viruses, protozoa, parasites, and worms	chlorine disinfection; ozone disinfection; ultraviolet radiation disinfection; filtration
dissolved solids	salts, small inorganic compounds	membrane filtration; chemical treatment; ion exchange; carbon adsorption
volatile organic compounds	small organic compounds that evaporate in the air	carbon adsorption; oxidation
odors	gases and volatile organic compounds	chemical scrubbers; carbon adsorption; biological filters

(Appendix E provides a description of the technologies listed in the preceding table.)

facility runs traditional wastewater treatment or wastewater-to-energy processes.

Wastewater treatment methods offer the following advantages: (1) easy to install and use; (2) small energy demand by most methods; and (3) inexpensive, except for ozone disinfection, ultraviolet disinfection, and ion exchange.

CARBON ADSORPTION

Adsorption is the process of removing chemicals from a solution by accumulating the chemicals on a solid surface. Engineers refer to this as the transition from an aqueous environment to a solid environment. Large wastewater treatment plants have more than one method for removing hazards from wastewater; adsorption supplements these treatments because of its efficiency and ease of use. Sustainable wastewater treatment plants, small rural plants on a strict budget, and sustainable homes act as ideal places to use adsorption for cleaning wastewater.

Three types of adsorbents, substances that adsorb materials from water, work in treatment: activated carbon, synthetic long-chain compounds called *polymers*, and silica-based adsorbents. Polymers and silica-based compounds are expensive, but inexpensive activated carbon has been important in wastewater treatment for many years. Activated carbon consists of any material with high carbon content, which is treated to increase its capacity to draw contaminants from water. Also important, activated carbon holds onto the contaminants and does not release them back into the water.

Makers of activated carbon use any of the following high-carbon materials as precursors: almond, coconut, or walnut hulls; wood; bone; or coal. The activation process contains the following two steps: (1) heating the material to 700°F (371°C) to burn off hydrocarbons and make a material called char; and then (2) activating the char by exposing it to steam and carbon dioxide to create a porous structure. The extensive pore system in the carbon greatly increases surface area and thus makes the material much more efficient as an adsorbent. Activated carbon tends to contain the following pore sizes:

1. macropores—greater than 25 nanometers (nm)
2. mesopores—greater than 1 nm and less than 25 nm
3. micropores—less than 1 nm

Wastewater treatment systems use either of two types of activated carbon: (1) powdered activated carbon (PAC) that contains small particles

(continues)

(continued)

(less than 0.075 mm diameter), or (2) granular activated carbon (GAC) that contains larger particles of about 0.1–2.4 mm diameter. Smaller particle sizes also increase the adsorbing surface area compared with large particles.

Carbon adsorption works by exposing the solid surface to a flowing liquid. Contaminants drift toward the solid-liquid interface located adjacent to the solid adsorbent. Contaminant molecules diffuse through this static interface and enter the particles' pores where they attach to sites called available adsorption sites, which are spaces on the adsorbent's surface that are unclaimed and available to catch and hold onto a contaminant. Because water flows in and around the adsorbent's pores, the method is often called carbon filtration. A variety of chemical interactions create the force that holds a contaminant on the surface: charge, covalent bonding, hydrogen bonding, or van der Waals forces. Covalent bonds occur when two atoms share electrons, while van der Waals forces develop when adjacent atoms share a weak attraction to each other because of an electromagnetic field.

Carbon adsorption offers an inexpensive and easy-to-maintain water cleaning method for sustainable homes, especially for treating gray water. The carbon can be reactivated once it has filled up with contaminants and no longer has many available adsorption sites remaining. Reactivation uses the same process as that used to activate char, that is, heating. For home systems, homeowners can simply replace the activated carbon filter every six months to a year after installing it rather than reactivate a used filter.

ENERGY FROM WASTEWATER

Sustainable wastewater treatment involves the reuse of the process's by-products for the treatment plant's operations. A typical wastewater treatment plant offers several options for generating energy or carrying out other conservation measures. Carefully planned wastewater treatment can provide the following benefits: (1) the natural flow of water due to gravity can act as an energy source; (2) methane and hydrogen production from anaerobic digestion serve as energy sources for powering the facility; (3) the reactions inside the digester produce heat that can be rerouted to other biological processes; (4) the treatment plant's gray water may be reused for flushing toilets; and (5) nutrient-rich sludge exiting the digester serves for landscaping the facility's grounds or may be sent to local farmers. In summary, in sustainable wastewater treatment, everything is used and reused to the maximum.

The wastewater industry resembles other industries today in trying to meet stricter pollution requirements while reducing energy use. Water and wastewater treatment in the United States uses only 2 percent of the country's total amount of energy, but with concerns over fuel and energy costs the industry faces the need to build more sustainable operations. Electricity needs of wastewater treatment plants vary by the amount of incoming wastewater they treat each day and the concentration of waste in the water. The wastewater treatment engineers George Tchobanoglous, Franklin Burton, and H. David Stensel, authors of *Wastewater Engineering: Treatment and Reuse,* have estimated that wastewater treatment processes use electricity in approximate proportions shown in the table on page 176.

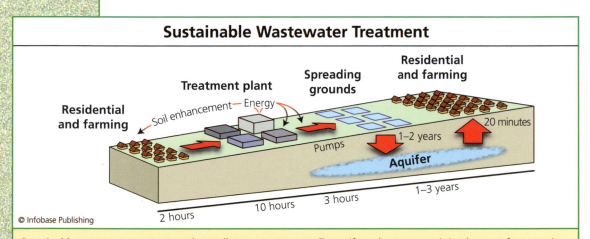

Sustainable Wastewater Treatment

© Infobase Publishing

Sustainable wastewater treatment plants allow water to naturally purify and consume minimal energy for pumping water. Methane gas from the treatment plant's anaerobic digester can provide some or all of this energy. The Earth purifies the water due to the soil's ability to remove particles and bind to many organic compounds.

TYPICAL ENERGY USAGE BY WASTEWATER TREATMENT PLANTS

PROCESS	PERCENT OF TOTAL PLANT ENERGY
sludge aeration	55.6
primary clarification and tank pump	10.3
heating	7.1
drying of solids	7.0
inflow pumping station	4.5
secondary clarification	3.7
process water	3.6
chlorine disinfectant mixing	3.1
lighting	2.2
sludge pump	1.6
inflow and outflow screens and filters	1.3

Sustainable practices in wastewater treatment provide some or all of the energy needs for running pumps for the aeration step and other pumps, heating, and lighting. Methane serves as the most efficient energy source for these purposes, helped to a smaller degree by hydrogen gas. Wastewater-to-energy plants recover methane in a three-step process that involves (1) dehydration, (2) gas cooling, and (3) removal of hydrocarbon compounds heavier than methane. This process results in pure methane of natural gas quality. Methane contains a heating value of 55.5 megajoules per kg of material compared with natural gas's 53.0 MJ/kg. Other fuels do not provide as much energy as methane: gasoline (48.1 MJ/kg); home heating oil (45.5); coal (28.5 MJ/kg); and wood (about 15 MJ/kg). Wastewater-to-energy therefore holds exciting promise for leading the way in new waste-to-energy technologies.

CONCLUSION

Water for drinking, irrigation, or industrial processes, as well as wastewater collection and treatment, has long been a major focus area in environmental engineering. Since delivery of clean water and removal of wastewater are so essential to health, this aspect of engineering developed early in human civilization. Today's mounting concern for the health of the environment has put increased pressure on safe wastewater management. To meet this challenge, the wastewater industry has made progress in developing better treatment methods that meld with sustainable methods.

Wastewater treatment has offered several opportunities to convert traditional treatment methods into self-sustaining treatment methods. Most of this opportunity comes from the by-products of bacterial digestion of wastewater sludge. The gases that these bacteria produce can be collected and converted into an energy source for generating heat or electricity. The newest wastewater treatment plants have already adopted this technology to meet part or all of their energy needs. Wastewater treatment has in fact led the way in the trend toward sustainable industry.

Sustainable wastewater treatment does not demand highly technical processes, so commercial wastewater treatment serves as a model for sustainable houses or other buildings. In addition to energy capture, builders of sustainable houses can follow wastewater management's lead on gray water reuse, natural digestion systems for waste, filtration, and adsorption using carbon filters. An engineer experienced in developing efficient waste-to-energy systems can suggest the most efficient methods of energy and water recycling and reuse.

Waste is a reality of society that will never disappear, so the waste and wastewater industries hold tremendous potential for finding better techniques of turning waste into an energy source. This serves the dual purpose of finding a use for the world's mounting waste problem and helping society move toward a more sustainable way of living and managing energy.

FUTURE NEEDS

Civilizations have always had individuals who were able to design a new tool or method to complete a job more efficiently. These early engineers began developing their technology when they used the first spear to bring down game and rubbed pieces of flint to make fire. Sometimes technology fills a need in society: A growing industrial population with increased desire to communicate spurred the invention of the telephone. Other inventions have become part of people's lifestyle because technology identified a need that people did not realize existed. The simple invention of Velcro, based on the spiny hooks on the surface of burrs, has been incorporated into hundreds of products. If there is any one theme that seems to hold true throughout the history of engineering, it is the constancy of engineering in people's lives: all things from a ballpoint pen to a massive industrial manufacturing plant are products of engineering.

A manufacturing plant also gives an example of how advanced human technology and engineering have become. Civilization has created structures so complicated that they bear little resemblance to anything found in nature. This was once lauded as one of civilization's greatest achievements, but people have come to learn that some behemoths of engineering have caused great harm to the environment. Today's environmental engineering might not be able to undo much of the damage the environment has suffered due to human development, but it can build a new sustainable future. It will do this not by subduing nature but by following nature's lead.

Environmental engineers have opportunities to reinvent how houses recycle their resources, how manufacturing plants manage energy and waste, and how an entire society travels yet minimizes fuel use and environmental destruction. The hardest part of this task comes not from technology but from the challenge of getting society to change its habits. Any new engineering feat that conserves the Earth's natural resources had better also offer convenience and low costs, or the public and industry might simply turn their backs on it. Environmental engineering will therefore not be isolated on an island; its successes will come because of cooperation from environmental organizations, the public, government, and, most important, from industries that will manufacture and use the new technologies.

Breakthroughs that contribute to sustainability and are accepted by society will likely no longer be the enormous engineering projects of the past. Huge projects such as China's Three Gorges Dam, for instance, have already been questioned as wasteful mistakes even before the construction has been completed. Environmental engineering will probably achieve success by mimicking the efficient form and function that already exists in nature. The idea of mimicking nature once seemed far outside the mainstream of design or architecture. In the immediate future, however, people may notice more and more structures built on the principles used in nature for creating strength, durability, and efficiency. The concept of less is more may become the guiding principle of almost all future structures conceived and designed by environmental engineers.

Appendix A

SUBJECT AREAS IN ENVIRONMENTAL ENGINEERING AND DESIGN EDUCATION	
MAIN SUBJECT AREA	**FOCUSED COURSEWORK**
Environmental Engineering	
computer science and mathematics	programming computer-aided design mathematical modeling computation
water systems	water sampling and measurement ecosystem hydrology water systems and landscaping groundwater watershed management fluid mechanics water treatment
natural sciences and environmental sciences	biology ecosystems ecology soil science solid waste management atmospheric science
energy	renewable energy management sustainable energy

Main Subject Area	Focused Coursework
Design	
architecture	design and structure theories
	mathematics
	geometry
	art and drawing
	architectural engineering
city and regional planning	urban studies
	housing in industrialized and developing countries
	transportation infrastructure
	land use
environmental science	ecology
	waste management
	geography
urban design	urban planning
	landscape design
	architecture
	computer analysis of structure
	visual technology

Appendix B

TYPES OF ENGINEERING	
ENGINEERING SPECIALTY	**INDUSTRIES SERVED**
aerospace	aerospace products and parts manufacturing
agricultural	food manufacturing
biomedical	medical equipment and supplies manufacturing
chemical	chemical manufacturing
civil	construction, architecture, urban design, transportation system design
computer hardware	computer and electronic product manufacturing
electrical	architecture, construction, technology
electronics	electronic product manufacturing, telecommunications
environmental	architecture, landscaping, water treatment, waste management, urban design
geological	materials science, mining, oil exploration
health and safety	state and local government
industrial	transportation equipment or machinery manufacturing

Engineering Specialty	Industries Served
marine and naval	ship and vessel manufacturing, government, marine studies
mechanical	vehicle, equipment, and electronics manufacturing, robotics
nuclear	nuclear power, life sciences
petroleum	oil and gas extraction

Source: U.S. Department of Labor

Appendix C

MILESTONES IN ENERGY USE	
Date	**Event**
10,000–9000 B.C.E.	humans discover fire and learn to harness its energy
5000	Phoenicians use wind power for ships
3000–2000	Egyptians collect and burn oil that floated to the top of ponds
3000–2000	Native Americans burn coal to make clay pots
3000–2000	Chinese collect natural gas for use as fuel
1769 C.E.	Nicolas-Joseph Cugnot invents steam-powered car
1792	William Murdock invents coal gaslight
1800s	emergence of machinery
1821	first U.S. natural gas well dug
1830	first hydropower dam put into service in Appleton, Wisconsin
1830s	coal-generated pollution accumulates in many cities
1840s	railroad lines expand across the United States
1859	first U.S. oil well was dug

Date	Event
1876	Alexander Graham Bell develops telephone
1879	Thomas Edison develops incandescent electric lightbulb
1880s	Butte, Montana's copper industry grows to meet need for new telephone wires
1885	Gottlieb Daimler invents gasoline-powered car
1888	first large windmill to generate electricity
1892	first gasoline-powered car
1913	Henry Ford develops mass production
1900s	coal and gas companies develop industry cooperation to supply energy to homes rather than homes generating their own energy
1900s	indoor plumbing and hot and cold running water moves into homes
1941	fission of uranium-235
1951	first nuclear power plant for making electricity
1953	first solar cell to produce electricity
1960	first commercial use of nuclear power
1963	Glen Canyon Dam completed on the Colorado River
1997	Introduction of first commercial hybrid vehicle (Toyota Prius)
2002	Japan installs solar panels on 25,000 rooftops
2009	Three Gorges Dam becomes fully operational

Appendix D

DEVICES USED FOR STORING ENERGY	
DEVICE	**DESCRIPTION**
lead-acid battery	rechargeable battery, often used in cars, that creates charge in a sulfuric acid solution between two oppositely charged lead electrodes, the anode and the cathode
zinc-air battery	non-rechargeable battery that creates charge from the oxidation of zinc with oxygen from air (replaces older mercury-air batteries)
flow batteries	large rechargeable energy-storage units in which the reactants, two different electrolyte systems, are stored outside the battery until needed
lithium ion battery	rechargeable battery in which lithium ions carry charge between the anode and the cathode
capacitor	stores electricity in two oppositely charged metal plates separated by a nonconducting substance; generates electricity by applying an outside current
flywheel	holds potential energy in a rotor that has been accelerated to high speed and then restrained; releases energy as the rotor is slowly released
compressed air	air is mechanically compressed to store high-level energy as heat
water (hydro pump)	water pumped to elevated reservoir stores potential energy, which it releases as water is released

Source: URL: http://electricitystorage.org/tech/technologies_technologies.htm

Appendix E

WASTEWATER TREATMENT METHODS	
METHOD	**DESCRIPTION**
aerobic/anaerobic bacterial digestion	bacteria degrade wastes in their normal metabolism
biological filters and reactions	specific microbial devices that remove contaminants from water, example: biofilms
carbon adsorption	adherence of contaminants to the surface of carbon granules
chemical oxidation	reaction that neutralizes a hazardous chemical by removing electrons
chemical precipitation	substances form aggregates with contaminants that cause the complex to settle out of water by gravity
chemical scrubbers	devices that contain filters and absorptive material that remove chemicals from vapors
disinfection	chemical or physical killing of microorganisms
filtration and screens	physical removal of solids by passing water through pores in a barrier
ion exchange	a substance removes metals or salts from water by exchanging with a harmless element
settling, clarification tanks	suspended constituents settle out of water over time by gravity

Glossary

ADVANCED TRANSIT (light transit)   a comprehensive urban transportation system designed specifically to save energy and fuel.

AERODYNAMICS   features related to reducing wind resistance in a moving vehicle.

ALTERNATIVE FUEL   any fuel energy source that reduces or eliminates the use of fossil fuel.

BIODIVERSITY   variety of different species, genes within a species, or different ecosystems.

BIOMASS   organic matter from plants, animal wastes, or wastewater treatment that can be used as fuel.

BIOMIMETICS   use of engineering principles to create biology-based structures.

BIOMIMICRY   basing human-made designs on the form and function of natural designs.

BROWNFIELDS   abandoned or idle industrial or commercial sites where redevelopment is hampered by hazardous contamination.

CARBON FOOTPRINT   measure of human activities' effects on the environment in terms of greenhouse gases produced and fossil fuels consumed.

CAR CULTURE   human behavior in which almost all activities require or desire the use of a car.

CLEAN COAL (low-sulfur coal)   coal treated to remove some sulfur compounds, mostly sulfur dioxide, for the purpose of reducing sulfur emissions when coal burns.

CLEAN SHIPS   oceangoing ships designed and operated to conserve fuel and energy and use renewable energy resources.

COAL GASIFICATION   conversion of solid coal to synthetic natural gas.

COMBUSTION   process in which oxygen combines with other molecules to form new compounds and release energy as heat.

COMPACT CITIES   city centers that add taller buildings as population grows rather than build outward toward suburbs.

CONCEPT CAR   an innovative car model that is not ready for public roads but incorporates new technologies that may be used in the future.

CONSTRUCTED WETLAND   human-made wetland based on the structure and vegetation of natural wetlands and used for the purpose of cleaning contaminants out of water.

DAYLIGHTING   use of natural light and sunlight to light the indoors.

DEPOLARIZATION   a condition in which positively charged particles and negatively charged particles become interspersed rather than separated at oppositely charged poles.

DISPERSED CITIES   cities that expand outward into suburbs and rural areas as population grows.

DISTRIBUTION CHAIN   the schedules, routes, storage sites, and delivery methods for transporting products from manufacturers to customers.

DRAFTING   the craft of drawing an accurate and precise depiction of a structure to be built.

DRAG   the forces that work against the forward movement of a moving vehicle.

DRAG COEFFICIENT (Cd)   a unitless value that describes the extent of a vehicle's aerodynamics.

ECO-BUILDING   any structure planned, designed, and built with the intent of having little or no negative impact on the environment.

ECOLOGICAL ACCOUNTING   the determination of a community's consumption and waste production in relation to technologies that can replace resources and reduce waste.

ECOLOGICAL DESIGN (also known as eco-design)   the design of structures that blend with nature, make nature visible, and do not harm nature.

ECOSYSTEM   community of species interacting with one another and with the nonliving things in a certain area.

EDIBLE LANDSCAPE   a parcel of land designed for the planting and use of vegetables and fruit-producing plants and trees.

END-OF-THE-LINE   a type of activity in which corrections and changes are made after a structure has been built or a product has been manufactured.

ENERGY GRID (power grid)   infrastructure used to store energy, usually electricity, until that energy is needed by a community.

ENVIRONMENTAL BURDEN   waste's harm caused to the environment or the energy needed to dispose of it.

FORM AND FUNCTION   the relationship between the shape and size of a structure and how the structure works.

FRONT-END   the preproduction planning of a structure or product's design, function, and operation.

GAIN   the amount of solar energy that a structure collects and retains.

GEOTHERMAL   a type of energy source that utilizes the steam and pressure emitted from geysers, vents, or hot springs.

GREEN CAR   a car designed to cause little or no harm to the environment.

GREENHOUSES GASES   natural and human-made gases that become trapped in the atmosphere and promote global warming; mainly methane, carbon dioxide, ozone, hydrogen sulfide, nitrogen oxides, and water vapor.

HAZARDOUS WASTE   any solid, liquid, or contained gas that can catch fire, is corrosive to skin or metals, is unstable and can explode, or can release toxic fumes or chemicals.

HEAT ISLAND EFFECT   situation in which a city's many brick and concrete buildings hold heat and thus raise the outside temperature.

HYBRID VEHICLE   any vehicle that uses more than one type of energy source in converting fuel into motion.

HYDROCARBON   a long chainlike compound with a backbone of carbon molecules and hydrogen molecules attached to each carbon; the energy-storage compound in petroleum.

INDUSTRIAL HYGIENE   occupation that deals with controlling workplace hazards.

INVASIVE SPECIES   plant or animal species that migrate into an ecosystem or are deliberately or accidentally brought in by people.

JOULE   the energy needed to lift an object weighing one Newton (0.445 pounds; 0.2 kg) 3.3 feet (1 m).

JUST-IN-TIME   type of manufacturing in which raw materials come to a factory in amounts needed for production, avoids warehousing and stockpiling of products and materials.

KINETIC ENERGY   energy contained in matter because of mass or velocity.

LEAN MANUFACTURING   planning and operations implemented for the purpose of reducing, but not necessarily eliminating, manufacturing wastes.

LIFT   the tendency of a moving car to catch air between it and the road.

MASS TRANSIT   buses, trains, and other forms of transportation that carry large numbers of people.

MEGAJOULE   a unit of energy equal to 1 million joules.

MICROBE (microorganism)   a microscopic and usually single-celled organism such as bacteria and protozoa.

NANOSCALE   describing an object that exists at the size of an atom or molecule.

NANOTECHNOLOGY   field of science that creates devices on a nanoscale, or the approximate size of atoms or molecules.

NATURAL CAPITAL (also known as natural resources)   Earth's natural materials and activities that sustain life and the economy.

NET ENERGY RATIO   usable energy derived from a process divided by the energy needed to produce the process's fuel.

OFF THE GRID   a type of activity or lifestyle that does not depend on a municipal power grid for energy because the activity or lifestyle makes its own energy.

OLD-GROWTH FORESTS   forests aged hundreds or thousands of years and developed from the natural succession of new vegetation on barren land.

PASSIVE ENERGY   energy generated by nonmechanized means without the need for any energy input.

PERMACULTURE   design and maintenance of ecosystems that grow an agricultural product.

POTABLE WATER   water treated to be safe for drinking.

POTENTIAL ENERGY   energy contained in an object because of its position; i.e., a boulder at the top of a cliff.

PROTOTYPE   a mock-up model of a vehicle or product used to illustrate the shape, size, and other design features.

QUAD   a unit of energy equal to 10^{15} Btu.

RADIATION   fast-moving particles, such as nuclear emissions, or waves of energy, such as sunlight.

RENEWABLE ENERGY   an energy resource than can be replenished rapidly (hours to decades) through natural processes, such as wind, tides, and solar energy.

RIBLET EFFECT   the reduction of drag in water due to tiny surface ridges called riblets that interact with the skin-water boundary layer.

RIBLETS   scalelike appendages, as found on a shark's skin, that reduce drag.

SCRUBBER   a filterlike device attached to a smokestack to remove hazardous particles and some gases.

SKIN FRICTION DRAG   resistance to movement due to a vehicle's surface texture and the surface's interactions with the air.

SUSTAINABLE   referring to a system's ability to survive for a period of time.

SUSTAINABLE LOOP   a structure's design features that enable the structure to reuse resources and generate energy.

SUSTAINABLE MANUFACTURING   activities that minimize natural resource consumption and waste production and maximize resource recycling in the making of a product.

THERMAL CAPACITANCE   capacity of materials to absorb and store heat for later use.

THERMAL ENERGY   energy in the form of heat.

THERMAL MASS   matter that stores heat.

THERMAL RESISTANCE   capacity of a material to block the flow of heat.

URBAN SPRAWL   expansion of cities horizontally into suburban and rural areas, leading to difficult commutes and traffic congestion.

VELOCITY GRADIENT   layers of flowing air or water in which each layer in sequence moves at a higher speed than the adjacent layer.

VIRTUAL COMMUTING   activity in which a person uses computer technology that creates the effect of the person being immersed in a three-dimensional workplace.

VIRTUAL REALITY   capability of computers to create a three-dimensional image that places the user in a lifelike setting.

VISIBLE SPECTRUM   portion of the electromagnetic, or light, spectrum that is visible to humans.

WASTE-TO-ENERGY   feature of a manufacturing or other type of facility that converts the energy in waste matter into a usable form of energy such as electricity.

WAVELENGTH   distance between repeating units of a wave, such as the peak to peak distance.

WATT   a unit of power equal to the work done at 1 joule (a standard unit of energy) per second or 1/746th horsepower.

WHITE MANUFACTURING (white biotechnology)   manufacturing that depends on biological processes and moderate temperatures rather than chemical processes and high heat.

WINGLETS   vertical fins at the end of airplane wings, making wings more aerodynamic and cutting fuel consumption.

XERISCAPING   method of landscaping that includes plant selection, irrigation, and shaping of the land for the purpose of minimizing water demand and water waste.

ZERO DISCHARGE   any operation that produces all of its own energy and recycles materials so that it puts no waste into the environment.

ZERO EMISSION VEHICLE   vehicle that produces no pollution.

ZERO ENERGY   type of structure design in which cooling, ventilation, and other activities require no energy consumption.

Further Resources

PRINT AND INTERNET

Air Transport Association of America. *ATA Airline Handbook.* Washington, D.C.: ATA, 2008. Available online. URL: www.airlines.org/products/AirlineHandbookTableofContents.htm. Accessed October 10, 2008. An online resource with excellent information on the challenges of today's airline industry, including environmental issues.

Amadei, Bernard. "Engineers Lend Technical Aid to Developing Countries." Interview by Spencer Michels. PBS Online NewsHour (12/7/07). Available online. URL: www.pbs.org/newshour/bb/science/july-dec07/engineers_12-07.html. Accessed January 29, 2009. This interview gives insight into the goals of Engineers Without Borders by the head of the U.S. chapter of the organization.

American Public Transportation Association. *2008 Public Transportation Fact Book,* 59th ed. Washington, D.C.: APTA, 2008. Available online. URL: www.apta.com/research/stats/factbook/documents08/2008_fact_book_final_part_1.pdf. Accessed November 25, 2008. Detailed resource on all sectors of transportation with extensive data tables.

Autodesk, Inc. "Playa Viva Strives for Sustainability." 2008. Available online. URL: http://usa.autodesk.com/adsk/servlet/item?siteID=123112&id=11761234. Accessed November 25, 2008. A short article describing how CAD helped design a sustainable community in Mexico.

Bash, Dana, and Suzanne Malveaux. "Bush Has Plan to End Oil 'Addiction.'" CNN.com (2/1/06). Available online. URL: www.cnn.com/2006/POLITICS/01/31/bush.sotu/. Accessed October 10, 2008. CNN covers President Bush's famed speech on the U.S. oil economy.

Bello, Marisol. "Ridership on Mass Transit Breaks Records." *USA Today* (6/1/08). Available online. URL: www.usatoday.com/news/nation/2008-06-01-mass-transit_N.htm. Accessed November 25, 2008. An article on the effects of U.S. gas prices on mass transit use.

Binns, Corey. "Secret to Abalone Shell Strength Revealed." LiveScience.com (7/16/07). Available online. URL: www.livescience.com/animals/070716_

strong_nacre.html. Accessed November 25, 2008. A brief article explaining the unique properties of natural abalone shell material.

Broder, John M., and Micheline Maynard. "As Political Winds Shift, Detroit Charts New Course." *New York Times* (5/20/09). Available online. URL: www.nytimes.com/2009/05/20/business/energy-environment/20emit. html?_r=1&scp=1&sq=&st=nyt. Accessed July 15, 2009. An analysis of the auto industry's responsibility to new fuel economy standards.

Capra, Frtjof, and Gunrer Pauli, eds. *Steering Business Toward Sustainability*. Tokyo: United Nations University Press, 1995. Available online. URL: www.unu.edu/unupress/unupbooks/uu16se/uu16se00.htm. Accessed November 1, 2008. Though this book was published more than a decade ago, it provides expert opinions and discussions that hold true in today's business world.

Cincotta, Richard P. "Note on Mound Architecture of the Black-Tailed Prairie Dog." *Great Basin Naturalist* 49, no. 4 (1989): 621–623. A detailed technical article on prairie dog burrow construction.

Coalition for Sustainable Transportation. "California Passes Complete Streets Law." Press release (4/29/08). Available online. URL: http://coast-santa barbara.org/2008/04/pedestrians-on-the-hill/#more-85. Accessed November 14, 2008. A short piece on a new law for roadways that encourage sustainable lifestyles.

CSX Corporation. "CSX Transportation's Lisa Mancini discusses Rail Benefits at Passenger Train Conference." Press release (10/21/08). Available online. URL: www.csx.com/?fuseaction=about.news_detail&i=49895. Accessed November 25, 2008. A company press release gives insight into the railroad industry's vision for its future.

Dannheisser, Ralph. "'Cow Power' Program Converts Animal Waste into Electricity." (2/21/08). Available online. URL: www.america.gov/st/env-english/2008/February/20080221103802ndyblehs0.5918238.html. Accessed November 11, 2008. An article explaining the rapidly emerging technology of methane waste-to-energy.

Davis, Mackenzie L., and David A. Cornwell. *Introduction to Environmental Engineering*, 4th ed. New York: McGraw Hill, 2008. A textbook providing excellent background on water systems engineering, waste management, pollution, and materials science.

Douglass, Elizabeth. "Sun-powered Homes Defy a Cool Housing Market." *Los Angeles Times* (9/25/07). Available online. URL: http://articles.latimes.com/2007/sep/25/business/fi-solar25. Accessed November 2, 2008. Presentation of a promising future for solar homes even during times of slow home sales.

European Union Road Federation. *Sustainable Roads.* Brussels: Brussels Program Center of the International Road Federation, 2007. Available online. URL: www.irfnet.eu/images/IRF_BPC_on_Sustainable_Roads_April_2007.pdf. Accessed October 10, 2008. A report on little-known considerations of road construction; one chapter specifically addresses effects on the environment.

Fan, Maureen. "Creating a Car Culture in China." *Washington Post* (1/21/08). Available online. URL: www.washingtonpost.com/wp-dyn/content/article/2008/01/20/AR2008012002388.html. Accessed November 25, 2008. The author traces the growth of car ownership in China.

Fields, Amber. "Fighting Cow Methane at the Source: Their Food." *Discover* (7/8/08). Available online. URL: http://discovermagazine.com/2008/aug/08-fighting-cow-methane-at-the-source. Accessed November 11, 2008. *Discover* provides a short, interesting article on alternative ways to tackle a greenhouse gas problem.

Foster, Lynn E. *Nanotechnology: Science, Innovation, and Opportunity.* Upper Saddle River, N.J.: Pearson Education, 2006. An overview of the current promise and concerns about nanotechnology with examples of future uses of nanoscale materials.

Gilliam, Harold. "Weather as Varied as the People." *San Francisco Chronicle* (7/9/01). Available online. URL: www.sfgate.com/cgi-bin/article.cgi?file=/chronicle/archive/2001/07/09/MN139536.DTL. Accessed November 7, 2008. An interesting explanation on the phenomenon of microclimates.

Grady, Sean M. *Virtual Reality.* New York: Facts On File, 2003. A clear presentation of new computer technologies that aid engineering, design, and transportation.

Habitats, Inc. "A Living Ecosystem." Available online. URL: www.habitatsinc.com/landscapes. Accessed November 6, 2008. Explanation of the main elements of eco-landscaping.

Hobstetter, David. "Daylighting and Productivity: A Study of the Effects of the Indoor Environment on Human Function." *Real Estate News and Articles* (March 2007). Available online. URL: www.thespaceplace.net/articles/hobstetter200703.htm. Accessed November 25, 2008. A well-written article on the many benefits of natural sunlight in schools, homes, and the workplace.

Howlett, Debbie, and Paul Overberg. "Think Your Commute Is Tough?" *USA Today* (11/30/04). Available online. URL: www.usatoday.com/news/nation/2004-11-29-commute_x.htm. Accessed November 25, 2008. An article that explains the reasons for long car commutes and their consequences on urban life.

Johnson, Caroline Y. "Zipcar Is Expected to Join with Rival Flexcar." *Boston Globe* (10/31/07). Available online. URL: www.boston.com/business/globe/ar-

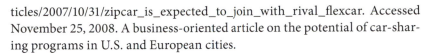

ticles/2007/10/31/zipcar_is_expected_to_join_with_rival_flexcar. Accessed November 25, 2008. A business-oriented article on the potential of car-sharing programs in U.S. and European cities.

Johnston, David, and Kim Master. *Green Remodeling: Changing the World One Room at a Time.* Gabriola Island, B.C., Canada: New Society Publishers, 2004. Useful information on alternative building materials and energy-efficient design.

Jordan, Robert Paul. "Our Growing Interstate Highway System." *National Geographic* (February 1968). An often-cited resource on the history of the U.S. interstate highways.

Kellert, Stephen R. "Social Ecologist and Author Stephen R. Kellert Shares His Views of Sustainable Design." *Sustainable Ways* 2, no. 1 (Autumn 2004): 1–4. Available online. URL: www.prescott.edu/academics/adp/programs/scd/sustainable_ways/documents/sw_autumn_20 04_sm.pdf. Accessed November 25, 2008. Insights from a scholar in the area of attitudes, nature, and ecological design.

Knuth, Marianne. "Kufunda Learning Village." Available online. URL: www.kufunda.org/people.php?sheet=2. Accessed November 10, 2008. A brief introduction to the mission and objectives of Zimbabwe's Kufunda Learning Village.

Lamb, Gregory M. "A Secret to Improving Cargo Ship Efficiency: Go Fly a Kite." *Christian Science Monitor* (4/3/08). Available online. URL: http://features.csmonitor.com/environment/2008/04/03/a-secret-to-improving-cargo-ship-efficiency-go-fly-a-kite. Accessed November 25, 2008. Description of a unique energy-saving device tried on an oceangoing ship.

Lawrence, Robyn Griggs. "Gold in the Hills." *Natural Home and Garden* (May/June 2006). One of many magazine articles available that provides ideas regarding building energy-efficient houses.

Lee, Ellen. "Crash May Virtually Change Commuting." *San Francisco Chronicle* (5/6/07). A description of the potential benefits of telecommuting and virtual commuting technology.

Levy, Claire. "Urban Sprawl Drives Up the Cost of Living." *Denver Post* (8/10/08). An article that explains how urban sprawl affects household economy.

Linn, Allison. "Long-distance Commuters' Road to Nowhere." MSNBC (7/23/08). Available online. URL: www.msnbc.msn.com/id/25722409. Accessed November 25, 2008. A sobering look at the factors that lead to long driving commutes.

Low, Nicholas, Brendan Green, Ray Green, and Darko Radović. *The Green City.* Sydney, Australia: UNSW Press, 2005. A book with clear details on sustainable buildings and towns, including a good amount of resources.

Marte, Jonnelle. "State Targets School Bus Pollution." *Boston Globe* (6/13/08). Available online. URL: www.boston.com/news/local/articles/2008/06/13/state_targets_school_bus_pollution/. Accessed November 14, 2008. An article that covers the attempts at cleaning up pollution from a local school bus fleet.

Millar, Andrea. "Kenya Jail Models Sustainable Wastewater Treatment." (4/23/08). Available online. URL: http://wiredberries.com/organic_living/2008/04/kenya_jail_models_sustainable.asp. Accessed November 9, 2008. A short article on alternative uses of wastewater treatment.

Millikin, Mike, and Alex Steffen. *Worldchanging: A User's Guide for the 21st Century.* New York: Harry N. Abrams, 2006. A wonderful resource for innovations underway and coming soon in all aspects of sustainable living.

Morris, John D. "Eisenhower Signs Road Bill; Weeks Allocates 1.1 Billion." *New York Times* (6/30/56). Description of the circumstances and some opinions regarding the birth of the U.S. interstate highway system; an interesting view of the times.

Motavelli, Jim. "Hybrid Hype." *E, The Environmental Magazine* (November/December 2008). An article explaining the odd business of building hybrid sport utility vehicles.

Mueller, Tom. "Biomimetics: Design by Nature." *National Geographic* (April 2008). An interesting set of examples of natural designs and the inventions that have or may come from them.

Nelson, Bryn. "Smart Appliances Learning to Save Power Grid." MSNBC (11/26/07). Available online. URL: www.msnbc.msn.com/id/21760974/. Accessed November 3, 2008. A news story on power-saving smart appliances.

Nelson, Erik N. "Locomotive Scrubber Debuts in Oakland Union Pacific Touts Rail Yard." *Oakland Tribune* (12/16/06). Available online. URL: http://findarticles.com/p/articles/mi_qn4176/is_/ai_n16900672. Accessed November 24, 2008. A press release that describes the railroad industry's efforts at making low-emissions locomotives.

Nolte, Carl. "Hybrid Boat Waves Hello." *San Francisco Chronicle* (5/17/08). A short article describing new hybrid technology for cleaner running recreational boats.

Otterpohl, Ralf. "Decentralized Wastewater Treatment (DEWAT)." Available online. URL: www.eawag.ch/organisation/abteilungen/sandec/schwerpunkte/ewm/dewat/index_EN. Accessed November 10, 2008. An article presenting issues to consider when planning wastewater treatment in developing countries.

Owen, James. "Clean Coal? New Technology Buries Greenhouse Emissions." National Geographic News (5/2/06). Available online. URL: http://news.

nationalgeographic.com/news/2006/05/0502_060502_coal.html. Accessed October 26, 2008. An article containing expert opinions on new less polluting coal technology.

Pacific Gas and Electric Company. "PG&E and Bioenergy Solutions Turn the Valve on California's First 'Cow Power' Project." News release (3/14/08). Available online. URL: www.pge.com/about/news/mediarelations/newsreleases/q1_2008/080304.shtml. Accessed November 11, 2008. An article with background on an emerging technology in alternative energy, cattle-produced methane.

Paumgarten, Nick. "There and Back Again." *New Yorker* (4/16/07). Available online. URL: www.newyorker.com/reporting/2007/04/16/070416fa_fact_paumgarten?currentPage=1. Accessed October 31, 2008. An engaging article providing insight on commuter-based society.

Petit, Charles. "Material as Tough as Steel? Abalone Fits the Bill." *New York Times* (3/22/05). Available online. URL: www.nytimes.com/2005/03/22/science/22abal.html?_r=1&oref=slogin. Accessed November 25, 2008. A science article that describe the structure of abalone shell and the reasons behind the shell's physical strength.

Philadelphia, Desa. "A $29 Million Green Parking Garage." CNNMoney.com (10/2/08). Available online. URL: http://money.cnn.com/2008/09/29/smallbusiness/green_parking.fsb/index.htm. Accessed November 25, 2008. A brief article describes how parking garages can support sustainability.

Prandtl, Ludwig. "On the Motion of Fluids with Very Little Friction." Paper presented at the Third International Mathematics Congress, Heidelberg, Germany, 1904. In Anderson, John D. "Ludwig Prandtl's Boundary Layer." *Physics Today* 58, no. 1 (2005): 42–48. Available online. URL: www.aps.org/units/dfd/resources/upload/prandtl_vol58no12p42_48.pdf. Accessed October 26, 2008. A technical article on a complex theory in physics in clear language and with excellent diagrams, in addition to interesting references to the history of physics.

Public Broadcasting System. "Secret Garden." Posted March 1998. Available online. URL: www.pbs.org/wnet/nature/episodes/secret-garden/introduction/3043. Accessed November 6, 2008. Text that accompanies an interesting video on the biology in backyard gardens.

Rich, Sarah. "Real-Time Energy Feedback Technology." WorldChanging.com (1/25/07). Available online. URL: www.worldchanging.com/archives/005903.html. Accessed November 5, 2008. A short article on energy usage feedback devices on a Web site that provides excellent resources on ecological living.

Ryker, Lori. *Off the Grid.* Layton, Utah: Gibbs Smith, 2005. A well-illustrated book that offers tips on building materials and sustainable heat and water use by examining actual houses.

ScienceWatch.com. "An Interview with Reid Ewing." (March 2006). Available online. URL: www.in-cites.com/papers/ReidEwing.html. Accessed November 25, 2008. A short interview containing interesting thoughts from an expert on the health effects of urban sprawl.

Seewer, John, and Doug Whiteman. "Edible Landscaping Saves Money, Homeowners Find." *Christian Science Monitor* (9/1/08). Available online. URL: http://features.csmonitor.com/gardening/2008/09/01/edible-landscaping-saves-money-homeowners-find/. Accessed November 13, 2008. One of many good articles available today with tips for planting edible gardens or landscapes.

Simborg, Mark. "Tapping into Nature's Genius." *San Francisco Chronicle* magazine (3/23/08). An article that gives insight into the thinking of an inventor who studies nature's designs.

Snell, Marilyn Berlin. "Can Coal Be Clean? New Ways to Burn a Dirty Old Fuel." *Sierra* (January/February 2007). Available online. URL: www.sierraclub.org/sierra/200701/coal.asp. Accessed October 27, 2008. Clear descriptions on new technologies in coal-fired power plants.

Stauffer, Nancy. "MIT Recommends Steps to Slash Gasoline Use by 2035." *MIT Energy Initiative Spotlight* (2008). Available online. URL: http://web.mit.edu/mitei/research/spotlights/slash-gas.html. Accessed November 25, 2008. Interesting short article on the fuel-usage crisis in the United States.

Stenzel, Volkmar. "'Riblet Effect' to Improve Aerodynamics of Vehicles and Aircraft." In What's Next in Science and Technology Podcast, (12/17/06). Available online. URL: www.whatsnextnetwork.com/technology/index.php/2006/12/17/p4211. Accessed November 25, 2008. A short article on new surface coatings based on natural structures, in this case, sharkskin.

Stodolsky, F., A. Vyas, R. Cuenca, and L. Gaines. "Life-Cycle Energy Savings Potential from Aluminum-Intensive Vehicles." Paper presented at 1995 Total Life Cycle Conference and Exposition, Vienna, Austria, October 16–19, 1995. Available online. URL: www.transportation.anl.gov/pdfs/TA/106.pdf. Accessed October 27, 2008. A detailed report, dated but remains an important resource for determining the energy costs of building a car.

Syphers, Geof. "Beyond Green." *San Francisco Chronicle* (3/25/07). A description of building a sustainable community in northern California.

Taylor, Michael. "Gearheads in Green." *San Francisco Chronicle* (5/20/07). An article providing background on the movement to create eco-friendly cars.

Tchobanoglous, George, Franklin L. Burton, and H. David Stensel. *Wastewater Engineering: Treatment and Reuse,* 4th ed. New Delhi: Tata McGraw-Hill, 2003. A detailed resource with descriptions of every aspect of wastewater treatment.

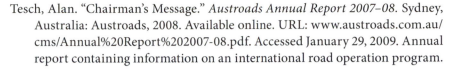

Tesch, Alan. "Chairman's Message." *Austroads Annual Report 2007–08*. Sydney, Australia: Austroads, 2008. Available online. URL: www.austroads.com.au/cms/Annual%20Report%202007-08.pdf. Accessed January 29, 2009. Annual report containing information on an international road operation program.

University of Florida. *Basic Principles of Landscape Design*. Available online. URL: http://edis.ifas.ufl.edu/MG086. Accessed November 9, 2008. A booklet presenting a clear description of the art and design elements of landscape design.

U.S. Department of Energy. *Nanotechnology: Energizing Our Future*. Seminars presented at Hot Topics in Science and Technology, Washington, D.C. (8/10/05). Available online. URL: www.sc.doe.gov/bes/presentations/archives_10AUG05.html. Accessed November 5, 2008. A series of lecture slides that presents a superb overview of nanotechnology in energy science.

U.S. Environmental Protection Agency. Office of Transportation and Air Quality. *Greenhouse Gas Emissions from the U.S. Transportation Sector*. Fairfax, Va.: EPA, OTAQ (2006). Available online. URL: www.epa.gov/otaq/climate/420r06003.pdf. Accessed January 27, 2009. A comprehensive report on all modes of U.S. transportation and the environment.

———. ICF International. *Energy Trends in Selected Manufacturing Sectors: Opportunities and Challenges for Environmentally Preferable Energy Outcomes* (March 2007). Available online. URL: www.ase.org/uploaded_files/efficiency_news/2007-05/epa_industrial.pdf. Accessed October 27, 2008. A government report covering the current status of energy consumption by industries and discussion of energy-saving alternatives.

———. Water Efficiency Program. *Water-Efficient Landscaping*. Available online. URL: www.epa.gov/WaterSense/docs/water-efficient_landscaping_508.pdf. Accessed November 7, 2008. A helpful booklet that describes water conservation in landscaping.

Vincent, Roger. "The Greening of Work." *Los Angeles Times* (8/27/06). A very good article that summarizes the new trends and health benefits of building ecologically friendly workplaces.

Vogel, Steven, Charles P. Ellington, and Delbert L. Kilgore. "Wind-Induced Ventilation of the Burrow of the Prairie-Dog, *Cynomys ludovicianus*." *Journal of Comparative Physiology* 85 (1973): 1–14. A technical examination of prairie dog burrow construction.

Watson, Traci. "Ship Pollution Clouds USA's Skies." *USA Today* (8/30/04). Available online. URL: www.usatoday.com/news/nation/2004-08-30-ship-pollution_x.htm. Accessed November 25, 2008. Discusses an overlooked aspect of pollution: oceangoing ships.

World Bank. "Greening Industry: New Roles for Communities, Markets and Governments." *Greening Industry* (October 1999). Available online. URL:

http://go.worldbank.org/DQLNZ0Y6R0. Accessed October 27, 2008. A brief synopsis of a lengthy World Bank report on making manufacturing less wasteful and more sustainable.

Yale University. Office of Public Affairs. "Researchers Design Artificial Cells That Could Power Medical Implants." *Science and Engineering News, Yale Bulletin* (10/9/08). Available online. URL: http://opa.yale.edu/news/article. aspx?id=6119. Accessed November 3, 2008. An article that is often cited as a good example of design by studying nature.

Zola, Zoka. "Architect Zoka Zola: From City Zoning to Modern Living." Video on BusinessPOV.com (2/14/07). Available online. URL: www.businesspov. com/article/141. Accessed November 25, 2008. A short video that follows architect Zola through her famous Chicago zero-energy house.

WEB SITES

American Society of Civil Engineers. Available online. URL: www.asce.org/asce. cfm. Accessed November 25, 2008. This is a comprehensive resource for the civil engineering profession.

Bureau of Transportation Statistics. U.S. Department of Transportation. Available online. URL: www.bts.gov. Accessed January 25, 2009. A detailed resource on all sectors of U.S. transportation.

Commonwealth Scientific and Industrial Research Organization (CSIRO). Available online. URL: www.csiro.au. Accessed January 25, 2009. CSIRO is Australia's resource organization for the country's industries, including energy and sustainable programs.

EarthSave. Available online. URL: www.earthsave.org. Accessed November 11, 2008. Good resource for topical articles on ecology and environmental science.

Energy Information Administration. Available online. URL: www.eia.doe.gov. Accessed November 1, 2008. An organization site, which is part of the U.S. Department of Energy, is a quality resource for all types of energy generation.

Energy Research Center of the Netherlands. Available online. URL: www.ecn. nl/en. Accessed October 27, 2008. Describes new technologies for energy conservation as it applies to industrial processes.

Energy Star. U.S. Environmental Protection Agency. U.S. Department of Energy. Available online. URL: www.energystar.gov/index.cfm?c=home.index. Accessed January 28, 2009. Contains energy calculator for finding options in home energy savings.

Engineers Without Borders-International. Available online. URL: www.ewb-international.org. Accessed January 29, 2009. Insight into engineering projects taking place in developing countries.

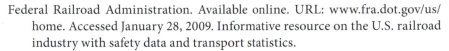

Federal Railroad Administration. Available online. URL: www.fra.dot.gov/us/ home. Accessed January 28, 2009. Informative resource on the U.S. railroad industry with safety data and transport statistics.

Green Car Congress. Available online. URL: www.greencarcongress.com. Accessed December 1, 2008. An excellent resource for cutting-edge vehicle technology, including transportation other than cars.

Institute for Transportation and Development Policy. Available online. URL: http://itdp.pmhclients.com/index.php. Accessed November 25, 2008. This organization works to promote sustainable transportation.

National Resources Defense Council. Available online. URL: www.nrdc.org. Accessed January 25, 2009. A good resource on the effects of human activities on the environment.

Research and Innovative Technology Administration. Bureau of Transportation Statistics. Available online. URL: www.bts.gov. Accessed November 25, 2008. A resource for air and freight industry statistics, developed in cooperation with the U.S. Department of Transportation.

U.S. Department of Agriculture. USDA Plant Hardiness Zone Map. Available online. URL: www.usna.usda.gov/Hardzone/ushzmap.html. Accessed November 7, 2008. Serves as a valuable landscaping resource since 2004.

U.S. Department of Energy. Available online. URL: www.afdc.energy.gov/afdc/ fleets/index.html. Accessed November 25, 2008. Excellent resources on alternative fuels and energy.

World Business Council for Sustainable Development. Available online. URL: www.wbcsd.org/templates/TemplateWBCSD5/layout.asp?MenuID=1. Accessed November 25, 2008. An international resource for sustainability with useful case studies.

Index

Note: Page numbers in *italic* refer to illustrations. The letter *t* indicates a table.